A Gardeners Guide

Planting For Ontario's Pollinators and Predators

First edition July 2024

ISBN: 978-1-7383793-0-9

Table of Contents

01 Welcome

03 What are Insects?

07 Overview of Insect Orders

09 What is Metamorphosis?

11 Invasive Insects

13 Insect Benefits

15 Beneficial Insects

19 What is a Native Plant?

21 Guide to Non-Native Plants

23 Plant Hardiness Zones

25 Cultivating Conservation

29 Building a Wildlife Garden

Insect Orders

33 Coleoptera

39 Beetles in the Garden
41 Morphological Marvels
43 Positive and Negatives
45 Discover Coleoptera
53 Planting for Coleoptera

61 Diptera

65 Understanding Diptera
67 Why do we need Diptera?
69 Morphological Marvels
71 Positive and Negatives
73 Discover Diptera
81 Planting for Diptera

89 Hemiptera

93 Unveiling Natures True Bugs
95 Why do we need Hemiptera?
97 Morphological Marvels
99 Positive and Negatives
101 Discover Hemiptera
109 Planting for Hemiptera

107 Hymenoptera

121 Why Hymenoptera Matter
123 Why do we need Hymenoptera?
125 Morphological Marvels
127 Positive and Negatives
129 Discover Hymenoptera
137 Planting for Hymenoptera

145 Lepidoptera

151 The Importance of Lepidoptera in Gardens
153 Morphological Marvels
155 Positive and Negatives
157 Discover Lepidoptera
165 Planting for Lepidoptera

173 Neruoptera

179 Why do we need Neruoptera?
181 Morphological Marvels
183 Positive and Negatives
185 Discover Neruoptera
193 Planting for Neruoptera

201 Odonata

205 Understanding Odonata
207 Why do we need Odonata?
209 Morphological Marvels
211 Positive and Negatives
213 Discover Odonata
221 Planting for Odonata

229 Orthoptera

233 What are Orthoptera?
235 Why do we need Orthoptera?
237 Morphological Marvels
239 Positive and Negatives
241 Discover Orthoptera
249 Planting for Orthoptera

257 Conclusion

Welcome

◇◇◇◇◇◇◇◇◇◇◇◇◇◇◇

Embark on a captivating journey through the unique realm of Ontario's native flora and fauna. Discover the intricate connections between these tiny creatures and our precious ecosystems. Immerse yourself in their fascinating world and uncover the pivotal roles played by insects and native plants in our ecosystems. North America is home to a remarkable diversity of over 60,000 insect species, some of our planet's oldest and most widespread creatures. In an average backyard, a staggering one thousand insects can coexist at any given time, highlighting the magnitude of their presence.

Insects comprise eighty percent of all animal species around the world, a true natural marvel. Their crucial contributions are vividly demonstrated through their irreplaceable role in ensuring the survival of flowering plants, such as the Cup plant (*Silphium perfoliatum*) and New Jersey tea (*Ceanothus americanus*), through pollination. In its natural balance, the Earth would revert to a state reminiscent of ten millennia ago if humans were to vanish abruptly. However, the disappearance of insects would throw the environment into disarray. Alarmingly, recent research has unveiled a rapid decline in insect populations in most areas, underscoring the pressing need for immediate and concerted conservation efforts.

Understanding that most insects are beneficial or harmless, despite troublesome insects dominating discussions, is essential. Over one million stunning species of insects have been identified worldwide, with an estimated over ten million species on Earth. Present on all seven continents, insects occupy a variety of environments and perform numerous ecological functions.

Consider this: UN Food and Agriculture Organization reports reveal that "75% - 95% of our crops" rely on pollinators. One in every three food items we consume directly results from pollinators' hard work. If we want to talk dollars and cents, pollinators add 217 billion dollars to the global economy. Honey bees alone are responsible for between 1.2 and 5.4 billion dollars in agricultural productivity in the United States. The benefits extend beyond food, with insects also producing valuable substances such as honey, wax, shellac, and silk, contributing significantly to our economy and enhancing our daily lives. This underscores the practical relevance of understanding and protecting these tiny yet mighty creatures and the significant impact their conservation can have on our lives and the economy.

In nature's balletic cycle, insects play soloist roles par excellence, from the gentle waltz of pollination to their essential role in breaking down plant material and dead animals. Predators like mantises (*Order Mantodea*) control other insect populations, thus forming a food source for larger animals, including mammals and birds. However, the documented declines in insect populations cast a shadow over this harmony due to habitat loss, climate extremities, and the omnipresent use of agrochemicals.

Yet, there is a glimmer of optimism. Research indicates that humans have significant potential to support insect life thriving in their gardens or confined areas. To cultivate insect-friendly spaces outdoors, refer to this comprehensive guide featuring Ontario's native wildflowers, grasses, and shrubs. These plants can attract beneficial insects and other wildlife, enriching our rural landscape.

All the plants highlighted in this guide are native to Ontario, and most species are available for direct purchase from Ontario Native Plants, an online refuge designed for Ontario gardeners. Their website offers over 100 species of native trees, shrubs, grasses, and flowers. Although this guide only showcases a fraction of the vast range of plants available through Ontario Native Plants, we recommend visiting their website to explore the additional extraordinary native plants suitable for your garden. Enjoy their user-friendly ordering system, which allows you to filter by your garden requirements. Ontario Native Plants was established in 2017 as an extension of a respected family-run wholesale nursery with over forty years of experience. Choose from having your order shipped or picked up if you are local. They are ready to help you enhance your yard's biodiversity and nourish the wildlife and insects within your property.

What Are Insects?

Insects, part of the expansive group of invertebrate animals known as arthropods, are distinguished by their jointed legs and internal limbs, allowing them to move within their hard outer shells. They possess a segmented body, a tough exoskeleton, three pairs of legs jointed at various points, compound eyes, and typically one or two pairs of wings. Their incredible diversity is evident through identifying over a million species on Earth, thriving in nearly all habitats. Insects are considered the most successful arthropods, dominating all animal groups within Insects. Their diverse forms have enabled them to inhabit almost every environment on Earth except cold polar regions, high-altitude uplands, and areas near active volcanoes.

Their ability to fly without spinal columns sets insects apart from other invertebrates. This feature has significantly contributed to their widespread distribution and colonization of new environments. Entomology, the branch dedicated to studying insects, is pursued by scientists known as entomologists. While primarily focusing on insects, entomologists also study related arachnids, such as spiders and scorpions.

In our garden landscapes, we commonly encounter a variety of insects, including butterflies, ants, bees, beetles, flies, crickets, mosquitoes, and more. These familiar creatures play crucial roles in ecosystems, serving as principal pollinators for flowering plants, primary food sources for insect-eating animals, and aiding in the decomposition of plants and animals. Apart from being essential pollinators, they serve as decomposers and are vital components of various food chains. Certain species significantly impact agriculture due to their pest status, while others are beneficial and used in biological pest control strategies and enhanced pollination services.

Eastern Cicada-killer Wasp,
Sphecius speciosus

By the Permian period, many modern insect Orders had established their primary body forms. For instance, Hymenoptera (*ants, bees, wasps, and sawflies*) and Lepidoptera (*butterflies and moths*) made relatively new appearances in the Jurassic rock records (210-145 million years ago). Fossilized specimens of Mantodea (*praying mantis*) have been found trapped in amber since the Eocene period (60-35 million years ago).

Insects represent the largest and most diverse group of animals on Earth, with over a million known species and many more being discovered each year. To manage this incredible diversity, scientists have categorized insects into various groups called "Orders." Each insect Order represents a distinct group with unique characteristics, behaviours, and evolutionary history.

Insects have adapted to diverse environments and have various feeding habits. They can be herbivores, carnivores, scavengers, or parasites, showcasing their remarkable versatility. Insects have existed for a long time, emerging before the Devonian period (400–360 million years ago) and mastering the art of flight by the Carboniferous period (360–285 million years ago). The strategic success of their flying adaptation led to their exceptional diversity during the Permian period (285–245 million years ago). No other group of arthropods achieved the ability to fly as insects did by that time.

Promethea Silkmoth,
Callosamia promethea

Red Flat Bark Beetle,
Cucujus clavipes

It's a fact that might surprise you: Within Ontario, there are over 10,200 individual species of insects, while Canada has over 17,000 individual species calling our country home. These insects, whether they crawl, glide, wiggle, or slither, are the unsung heroes of our ecosystem. They maintain healthy soil, recycle nutrients, pollinate flowers and crops, and control pests, all of which harmonize our environment. But it's not just nature that's working to protect insects. It's our everyday actions that can make a difference. By planting native gardens, avoiding pesticides, and conserving habitats, we are all contributing to the safeguarding of these vital creatures.

Now, let's delve into the fascinating world of insect Orders. Modern science has identified a staggering twenty-nine Orders, each with unique characteristics and species. These Orders are not just names and numbers, they represent a diverse and intricate web of life. In your everyday life in Ontario, you're most likely to encounter insects from the following eight Orders:

Coleoptera (*Beetles*) Coleoptera, commonly known as beetles, is the most diverse and abundant order of insects. One of the distinctive features of beetles is their hardened forewings, known as elytra, which cover and protect their delicate hindwings. This adaptation gives them a tough outer covering while allowing them to fly when necessary. Beetles exhibit an extraordinary diversity in size, shape, and habitat. Over 350,000 known species can be found in almost every terrestrial and freshwater habitat worldwide. Across Ontario there are over 2,000 individual species of beetles. This incredible diversity has enabled beetles to thrive in various environments, from tropical rainforests to arid deserts.

Diptera (*Flies*): Diptera, commonly known as flies, are a diverse order of insects with only a single pair of wings. The hind wings are modified into small, knob-like structures called halteres, distinguishing them from other winged insects. Diptera encompasses various species, including houseflies, mosquitoes, fruit flies, and hoverflies. Over 1,700 species of Diptera call Ontario home, while over 2,600 species call Canada home. Flies play crucial ecological roles as essential pollinators in many ecosystems. However, some species are considered pests and can have significant economic impacts, while others function as vectors of diseases, making them a concern for public health. These diverse characteristics and ecological roles make Diptera a highly significant and fascinating group of insects.

Hemiptera (*True Bugs*): Hemiptera, commonly known as true bugs, are a diverse group of insects with mouthparts that are adapted for piercing and sucking. This adaptation allows them to feed on plant sap or the blood of animals. Hemipterans are characterized by their distinctive triangular-shaped scutellum on their back and typically possess either fully membranous wings or partly hardened. This insect order includes various species, such as aphids, cicadas, and stink bugs, each with unique characteristics and behaviours. 1,100 plus species of Hemiptera call Ontario home.

Hymenoptera (*Bees, Wasps, and Ants*): Hymenopterans are a diverse group of insects known for their complex social structures and intricate behaviours. They exhibit various behaviours, including cooperative colony formation and complex communication systems. With two pairs of wings, these insects are known for their agility and flight capabilities, which are essential for their foraging and reproductive activities. Females of many hymenopteran species possess a specialized structure called an ovipositor, which serves multiple functions such as laying eggs, digging, or stinging. This adaptation allows them to navigate and manipulate their environment effectively. Over 1,400 Hymenoptera call Ontario home.

Lepidoptera (*Butterflies and Moths*): Lepidoptera, which includes butterflies and moths, are distinctive because their wings are covered in tiny scales. Butterflies are renowned for their brightly coloured and vividly patterned wings, while moths typically have more muted colours and more superficial wing structures. These creatures undergo complete metamorphosis, which involves distinct stages, including the larval (caterpillar) and pupal phases, before emerging as adults. Moths undergo pupation within cocoons, creating protective enclosures of silk, while butterflies form chrysalises during their pupal stage. There are over 3,100 species of Lepidoptrea within Ontario, with Moths being the higher population.

Neuroptera (*Lacewings, Antlions, and Owlflies*): Lacewings belong to the insect order Neuroptera and are characterized by their intricate, net-like wings. Their larvae are voracious predators, feeding on aphids and other harmful insects. This natural form of pest control helps to maintain ecological balance and reduce the reliance on chemical pesticides. In Ontario there are just over 40 species of Neuroptera.

Odonata (*Dragonflies and Damselflies*): Odonata, also known as dragonflies and damselflies, are predatory insects known for their large, transparent wings and long, slender bodies. They are highly skilled fliers commonly found close to freshwater habitats such as ponds, lakes, and rivers. Dragonflies typically have stouter bodies and hold their wings horizontally when at rest. At the same time, damselflies are characterized by their slender bodies and how they fold their wings vertically when not in flight. 170 Odonata species call Ontario home.

Orthoptera (*Grasshoppers, Crickets, and Katydids*): Orthopterans, which belong to the Orthoptera, are characterized by their powerful hind legs, which are specially adapted for jumping. They possess chewing mouthparts and are known for producing distinct sounds by rubbing their wings or legs together. Orthoptera includes various species, such as grasshoppers, crickets, katydids, and locusts. Just over 110 species of Orthoptera can be found across Ontario.

The insect world is prosperous and diverse, encompassing various sizes, shapes, and colours. While most insects pose no threat and benefit humans, certain species can cause significant harm. Identifying these harmful insects involves a deep understanding of their behaviour, preferred hosts, life cycles, and the potential extent of damage they can inflict. Fortunately, there are numerous proven methods for managing harmful insects, including cultural practices, careful selection of plant varieties, and the use of mechanical and biological controls, all of which have been highly effective in controlling and minimizing the damage caused by harmful insects.

Ontario, Canada, is home to a rich and diverse insect population that maintains the province's biodiversity. The various regions within Ontario provide a wide range of habitats for insect life, reflecting the province's diverse environments. From the tundra along the shores of Hudson Bay to the mixed forests of Lake Superior's west, the coniferous forests of the Canadian Shield, the deciduous forests of Southern Ontario, and the Carolinian woodlands in the south, Ontario's insect population thrives in a variety of ecosystems.

Overview of Insect Orders

◇◇◇◇◇◇◇◇◇◇◇◇◇◇◇◇

Knowing how to classify insects by their Order can be helpful as a home gardener. This can help you identify whether an insect is a beetle, wasp, or butterfly, giving you essential information about their life cycle and habitat. By understanding this standard classification system, you can better control pests and maintain a healthy garden. In this hierarchical system, organisms are grouped into five kingdoms, with insects belonging to the Animal Kingdom within the phylum Arthropoda. Arthropods, characterized by body segmentation and an exoskeleton, represent a substantial portion of known animal species. Several features separate insects from the other arthropod classes, dividing the body into three central regions (head, thorax and abdomen), three pairs of legs on the thorax and one pair of antennae. Insects, falling under the class Insecta, are further broken down into Orders, Families (e.g., *Aphidae, Muscidae, Blattidae*), and ultimately Genera and Species. Common names, such as Corn earworm or Tomato hornworm, may also be assigned to insects based on their habitat or behaviour. "Common names" often refer to large groups of insects, such as Families or Orders.

Additional groups, marked with super- (above) or sub- (below) prefixes, may fall between the listed groups. Superfamily groups are positioned between Order and Family, while Subfamily groups are between Family and Genus. A complete insect name, which includes the Genus, Species, and author names, follows strict taxonomy rules. The author's names are in parentheses if the classification has changed, highlighting the precision and accuracy of scientific names in insect classification. This is crucial to remember when using scientific names in your gardening practices.

Halloween Pennant,
Celithemis eponina

Terminology, Classification, and Use of Scientific Names

Let's delve into the world of insect names. Common names, like 'honey bee' or 'sawfly,' are written in lowercase letters unless they include a proper noun. Species within the classification are documented as two words, such as 'honey bee,' and as one word for those outside, like 'sawfly,' not in the Diptera Order. On the other hand, scientific names (*Genus, Species, and Subspecies*) are italicized or underlined, with the Genus always capitalized. If the classification has changed, the author's names are in parentheses, following. Understanding these naming conventions is a key step to becoming a knowledgeable home gardener.

Insects belong to the larger group Arthropoda, encompassing animals with segmented legs, bodies, and exoskeletons. The phylum Arthropoda includes spiders, ticks, mites, centipedes, millipedes, shrimps, lobsters, and more. Entomology focuses on two classes: Hexapoda or Insecta (insects) and Arachnida (*spiders, ticks, mites, scorpions, and relatives*). Some other arthropod classes, like Diplopoda (*millipedes*) and Chilopoda (*centipedes*), are considered by entomologists. Occasionally, Entomologists refer to non-arthropod groups like snails and slugs (*Phylum – Mollusca*).

The insect world is a complex web of life where each Order plays a unique role. From the precise pollination carried out by bees to the dynamic decomposition efforts of beetles, every Species has specialized adaptations and symbiotic relationships that sustain entire ecosystems. Understanding the significance of these roles is essential in fostering a deep appreciation for the sophistication and interdependence of the insect world.

Here is an example using the native Glacial lady beetle, *Hippodamia glacialis*:

Kingdom: Animalia
Phylum: Arthropoda
Class: Insecta
Order: Coleoptera
Family: Coccinellidae
Genus: Hippodamia
Species: H. glacialis

What Is Metamorphosis?

◇◇◇◇◇◇◇◇◇◇◇◇◇◇◇

Most insects start their life within an egg and have a growth phase before adulthood. During this non-reproductive phase, insects undergo regular moults (the process of shedding old skin and growing a new one) to allow for growth. Once they become adults, growth stops, and they can breed. Most adult insects also develop wings. Insects born with wings but do not pass through a pupal stage (a stage of metamorphosis in which an insect changes from a larva to an adult) are called hemimetabolous. They undergo an incomplete metamorphosis. Hemimetabolous insects include grasshoppers, crickets, true bugs, dragonflies, mantids, mayflies and stoneflies. Hemimetabolous larvae are active and easily mistaken for adult insects. They have well-developed legs, and their adult forms may look similar to their larval forms.

Metamorphosis, meaning change in form or structure, refers to all stages of development. The stage of life between each moult is called an instar. The number of instars and moults varies between species and depends on food supply, temperature, and moisture. The pupal stage is a period of transformation from larva to adult. It is a crucial stage where many tissues and structures are entirely broken down, and the adult's "true" legs, antennae, wings, and other structures are formed. The adult insect does not grow in the standard sense. The adult period is mainly one of reproduction and is sometimes of short duration. Their food is often entirely different from that of the larval stage.

Some insects that undergo incomplete metamorphosis are highly mobile and active from the moment they hatch. For example, a Bush cricket (*Family Tettigoniidae*) larva in its first instar (the stage between each moult) may be only a tenth of the size it will eventually attain. However, it is a near-perfect miniature of an adult, down to its elongated antennae and leaping hind legs. It has only tiny wing buds, and its reproductive organs have yet to develop fully. As it passes through successive moults, its abdomen and wing buds become more prominent, and (in females), the ovipositor (an organ used by some animals to lay eggs) becomes visible in later instars.

Some hemimetabolous insects hatch as a less developed prolarva stage. For example, the prolarva of a damselfly has no legs and no feeding mouthparts but is capable of limited movement. If the damselfly hatches out of the water, it will wriggle its way to water, which will moult to its long-legged, actively feeding second instar.

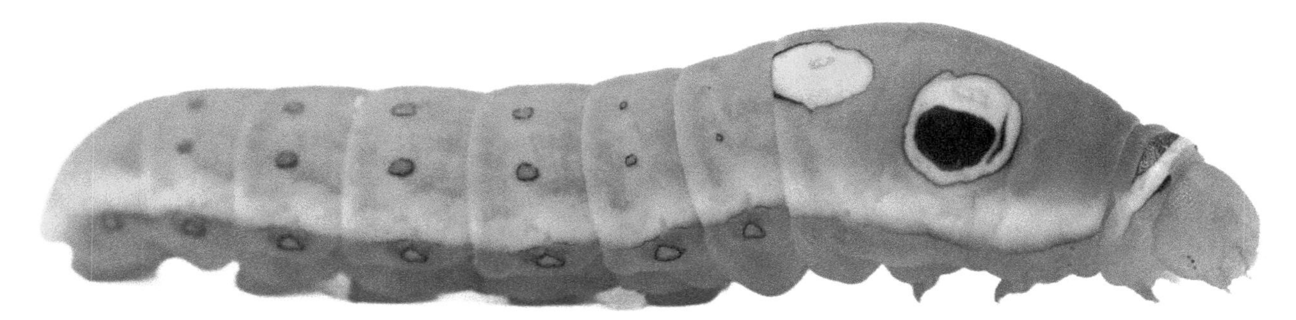

Spicebush Swallowtail (Caterpillar),
Papilio troilus

Diverse hormones regulate moulting. When the juvenile hormone levels drop in the final larval instar, the insect moults into the fully adult stage. The moulting process is the same as between larval instars, but in the case of aquatic insects, the larva usually fully exits the water first. After it has cracked through its larval cuticle, the newly emerged adult insect must transform its tiny, crumpled wings into functional flight-ready structures. It does this by swallowing air, thus increasing the pressure of the hemolymph in its thorax. This forces hemolymph into the wing veins, causing them to expand and stiffen.

The pupal development stage separates the holometabolous insects from the more primitive hemimetabolous species. When an insect larva is ready to pupate, it stops feeding and seeks a suitable site for pupation. Among Lepidoptera, for example, hawk moth larvae climb down from their food plant and seek a patch of soft ground, burying themselves in the earth. Some caterpillars spin a protective silk cocoon around themselves before pupation. Butterfly caterpillars often anchor themselves to a vertical plant stem or hang by their tail ends from a horizontal twig. Many Lepidoptera larvae use silk from their labial glands to stick themselves in position before pupating and spin a protective cocoon of silk within which they pupate. The cuticle of the newly formed pupa soon becomes firm. Those that are not camouflaged or hidden in some way risk being eaten by predators.

Eastern Black Carpenter Ant,
Camponotus pennsylvanicus

Invassive Insects
Unraveling Their Impact

◇◇◇◇◇◇◇◇◇◇◇◇◇◇◇

Non-native and invasive insects are subjects of prime importance in entomology because of their ecological intricacies, which require careful examination. These insects, also known as exotic or introduced species, are transported outside their natural habitats, often through human means such as international trade, tourism, or intentional introductions. Once in new habitats, they may inadvertently establish populations under unfamiliar ecological conditions and potentially exploit available resources.

The distinction between non-native and invasive insects is not just a matter of origin but also of potential harm to the ecosystems they inhabit. Invasive insects can displace native species, disrupt the ecological balance, and even lead to local flora and fauna extinction. Their success is often attributed to their adaptability to new environmental conditions, effective reproductive tactics, and the absence of natural predators in their new habitats.

Invasive non-native insects present a grave threat to the ecosystems where they are introduced. They breed rapidly and can disrupt the delicate balance that sustains biodiversity. For instance, invasive insects like the Spongy moth (*Lymantria dispar*) wield a voracious exploitative power over their Indigenous counterparts when feeding, potentially leading to the extinction of Indigenous plants and animals, and a significant loss of biodiversity.

The adaptability of invasive insects to different environmental conditions is critical to their survival in new habitats, where native organisms may struggle. These populations can rapidly expand due to their highly efficient reproduction strategies, intensifying their negative impacts on local ecosystems. Invasive insects can thrive by multiplying uncontrollably and causing disturbances within established ecological interactions, particularly in regions lacking natural enemies that might keep them in check. The crucial responsibility to track and control these alien species lies with us, as it is a vital step in protecting the diversity and stability of global ecosystems.

Spongy Moth,
Lymantria dispar

The presence of non-native and invasive insects can have far-reaching and complex ecological ramifications. They can disturb food webs, alter nutrient cycles, and cause imbalances within local ecosystems. They may harm native plant-animal communities through predation, competition for resources, or transmission of pathogens, thus posing a threat to biodiversity maintenance and ecosystem resilience. These invasive pests also have a significant economic impact on crops and forests, leading to substantial financial losses with increased management costs.

For effective mitigation measures against non-native and invasive insects, it is crucial to understand what factors make them successful invaders. No natural enemies or diseases frequently regulate their population sizes, leading to their limitless spread in the new environment. Moreover, having no historical co-evolution with the local flora gives them a competitive edge, which makes them more effective in resource exploitation.

Some of the methods Entomologists use in studying and controlling non-native/invasive species include ecological monitoring programs, chemical control, biological control, or physical (mechanical) control. Biocontrol involves using natural enemies like predators, parasites and pathogens to regulate the populations of invasive insects. However, this has to be governed by careful consideration and risk assessment so that the control agents do not turn out to be invasive themselves.

Studying insects that are not native to a particular environment and those that invade new areas helps us understand how organisms and their surroundings interact. This research also highlights the ecological disturbances caused by human activities. As a result, it is crucial to comprehend the mechanisms that speed up the increase in global trade and travel. This understanding is essential to safeguard biodiversity, preserve healthy ecosystems, and maintain global health.

European Mantis,
Mantis religiosa

Insect Benefits

Delving Into Their Value

◇◇◇◇◇◇◇◇◇◇◇◇◇◇

The balance of biodiversity in our gardens depends on how well insects can keep it. Apart from performing their role as pollinators, insects affect a garden ecosystem's overall health and well-being. Beneficial insects like native ladybirds and predatory beetles such as the Six-spotted tiger beetles (*Cicindela sexguttata*) help keep harmful insect populations in check by feeding on them, thus avoiding using chemicals. Bees are also vital pollinators for flowering plants, resulting in bountiful harvests. Insect activities such as decomposition and nutrient cycling help enhance soil fertility, too. Native plants like Harebell (*Campanula rotundifolia*), Spotted Joe Pye (*Eupatorium maculatum*), and Milkweed species such as Butterfly Milkweed (*Asclepias tuberosa*) should be included while creating an atmosphere where beneficial insects feel at home and have shelter but no harmful pesticides. By following these practices, we can assist in developing an eco-friendly environment that is good for gardening and nature in general.

Insects Contribute Significantly to Gardening in Several Ways:

Insects are crucial in pollinating blossoms, thus facilitating the production of fruits, seeds, vegetables, and flowers. Many common fruits and vegetables, such as melons and squash, rely on insects for adequate pollination. Ornamental plants, both indoors and outdoors, such as chrysanthemums, iris, orchids, and yucca, also benefit from insect pollination. Pollination brings us one in every three bites of food, a statistic that underscores insects' importance in food production. Without insects and pollinators, multiple studies by scientists have shown that humans would go hungry within months of their demise.

Insects are not just beneficial in our gardens; they are also our allies in agriculture. They play a multifaceted role, acting as natural weed controllers through processes like allelopathy. Allelopathy is a fascinating mechanism where an organism releases chemicals that affect the growth and survival of other organisms. It's like a plant releasing a chemical that prevents other plants from growing too close. These chemicals, known as allelochemicals, can positively and negatively impact the target organisms and the community. Beneficial insects, including parasitoid wasps and predatory beetles, target specific pests, reducing the need for chemical pesticides. Insects also bolster crop health by facilitating pollination, enhancing yields, and maintaining ecological balance. Their diverse feeding habits and allelopathic behaviours demonstrate how insects can positively impact agricultural ecosystems, promoting sustainability and stability in farming practices.

Two-marked Treehopper Complex,
Complex Enchenopa binotata

Insects contribute immensely to soil health by burrowing through the surface layer, enhancing soil aeration and drainage. Their movements create small tunnels in the soil, allowing air and water to penetrate deeper, which is crucial for plant roots. Their actions also promote nutrient cycling and microbial interactions, fostering overall fertility. Furthermore, insects' decomposed bodies and excrement serve as natural fertilizers, enriching the soil with essential nutrients and organic matter. This dual impact highlights insects' vital role in maintaining soil quality and sustaining healthy ecosystems.

Insects serve as essential scavengers and are pivotal in ecosystem sanitation by consuming dead animals and plants. Their efficient consumption accelerates the decomposition of carcasses and dung, facilitating nutrient recycling and maintaining ecological balance. Additionally, certain insects, like Dung beetles (*Order Coleoptera*), contribute to pest control by reducing the breeding sites for disease-carrying pests. This scavenging behaviour emphasizes the diverse ecological services provided by insects in waste management and ecosystem sustainability. Insects have diverse feeding habits. Some, like bees, feed on nectar and pollen, while others can be predators, like ladybugs, feed on aphids and other garden pests. This diversity of feeding habits is crucial in maintaining balance among insect populations and preventing any one species from overpowering the ecosystem.

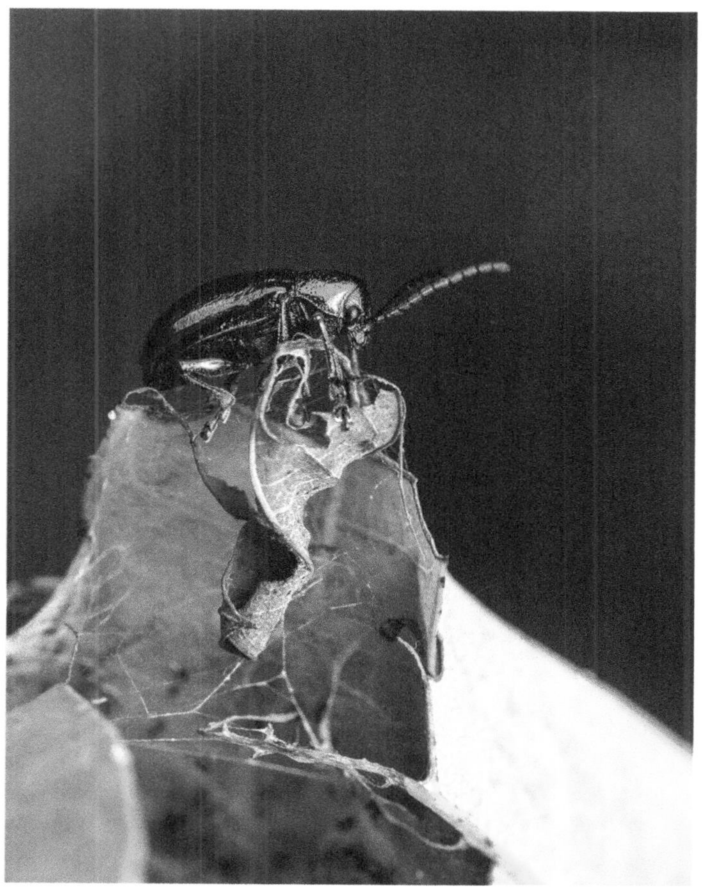

Right: Monarch,
Danaus plexippus

Left: Dogbane Leaf Beetle,
Chrysochus auratus

Benefical Insects
The Three Categories:

Predators capture and eat other organisms, such as insects or mites. Predators include Ladybugs (*Family Coccinellidae*), Ground beetles (*Family Carabidae*), Lacewings (*Family Chrysopidae*), and Yellowjacket (*Subfamily Vespinae*).

Parasitoids are insects that parasitize other insects. Their immature stages develop on or within their host, eventually killing it. Parasitoids may attack all stages of their host (eggs, larvae, nymphs, pupae, and adults). Examples include the Hornworm Parasitoid Wasp (*Cotesia congregata*).

Pollinators include honeybees, leafcutter bees, other wild bees, butterflies, moths and other insects that visit flowers to feed on nectar and pollen. Pollinators transfer pollen in and between flowers of the same species (pollination), which is essential to plant seed and fruit production.

Predators

Predatory insects, the unsung heroes of our ecosystems, play a pivotal role in natural pest control. Found in agriculture, forestry, and gardens across Ontario, these insects actively hunt and feed on harmful pests, maintaining a delicate balance in our ecosystem. Ladybugs and lacewings, for instance, are natural controllers of whiteflies, thrips, caterpillars, leaf miner larvae, and aphids.

Advantages of using predatory insects

Using predatory insects for pest control offers several advantages, promoting a sustainable and eco-friendly approach to managing pest-related issues. Here are some key benefits:

Sustainable pest control

Predatory insects provide a natural and non-toxic alternative to chemical pesticides. This reduces the environmental impact associated with synthetic chemicals, safeguarding the overall ecosystem.

Reduced need for chemicals

The use of predatory insects often results in a decreased reliance on chemical pesticides. This reduction benefits the environment and human health, as it minimizes exposure to potentially harmful chemicals and reduces the amount of synthetic chemicals released into the ecosystem.

Minimized resistance development

Unlike chemical pesticides, which pests can develop resistance to over time, predatory insects provide a dynamic and evolving solution. These resilient creatures are less likely to be outsmarted by pests, maintaining their effectiveness over the long term and giving you peace of mind in your pest control efforts.

Parasitoids

Insects that parasitize other insects. The immature stages of parasitoids develop on or within its host, eventually killing it. Parasitoids may attack all stages of their host (eggs, larvae, nymphs, pupae, adults).

Wasps and flies contain the vast majority of insect parasitoids. There are over 70,000 parasitoid species across the globe. The word "wasps" generally brings to mind Yellowjackets or Hornets, but most Wasp species are, in fact, parasitoids, ranging in shape and size from small 0.008-inch fairyflies (who are wasps) to the five-inch-long Megarhyssa wasps.

Parasitoids, with their fascinating life strategies, attack all life stages of arthropods. They employ various tactics to get their eggs on or in the host. Most wasps use their ovipositor to insert their eggs in or on their prey. On the other hand, parasitoid flies lack an ovipositor capable of piercing their host's exterior. Instead, they either glue their eggs onto the host or lay eggs on plants their host eats. Eggs ingested by the correct host insect then hatch in the host's gut. Parasitoids are incredibly specific to the life stage of hosts they attack. Even if other life stages of the host are present, the adult parasitoid will likely not even consider them as a potential host for her eggs.

Notable parasitoids in our area include:

- The Swift Feather-legged Fly (*Trichopoda pennipes*) attacks Squash bugs and Stink bugs.
- Hornworm Parasitoid Wasp (*Cotesia congregata*) attack tobacco hornworms (*Manduca sexta*)
- Tiphia wasps attack white grubs.

Parasitoids perform a vital ecosystem service by suppressing pest populations. The majority of pests are attacked by at least one parasitoid. There are three ways to improve pest suppression by parasitoids. The first is to limit insecticide use as much as possible. Insecticides are often toxic to many other beneficial insects. Using insecticides only when necessary based on scouting will greatly benefit parasitoids in your farm and garden.

Tomato hornworm, *Manduca quinquemaculata*

With braconid wasp cocoons, *Cotesia congregates*

Pollinators

A pollinator is anything that helps carry pollen from the male part of the flower (stamen) to the female part of the same or another flower (stigma). The pollen movement must occur for the plant to become fertilized and produce fruits, seeds, and young plants. Some plants are self-pollinating, while others may be fertilized by pollen carried by wind or water. Still, other flowers are pollinated by insects and animals - such as bees, wasps, moths, butterflies, birds, flies and small mammals, including bats.

Insects and other animals such as bats, beetles, and flies visit flowers for food, shelter, nest-building materials, and sometimes even mates. The sheer diversity of these pollinators, including many bee species, butterflies, birds, and bats, is a testament to the intricate web of life that supports our food production. This pollen movement from flower to flower is not just crucial; it's awe-inspiring.

Why are pollinators important?

Do you like to eat?

One out of every three bites of food you eat exists because of the efforts of pollinators. They are crucial not only for our food but also for the food and habitat of animals, making them an integral part of our ecosystem.

Do you like clean air?

Healthy ecosystems depend on pollinators. Insects and animals pollinate at least 75 percent of all the flowering plants on earth! This amounts to more than 1,200 food crops and 180,000 different types of plants—plants which help stabilize our soils, clean our air, supply oxygen, and support wildlife.

Do you want a healthy economy?

In Canada alone, honey bees' invaluable pollination service contributed over $3 billion to crop production in 2010, while pollination by other insect pollinators contributed nearly $10 billion. This staggering economic value underscores the urgent need to protect and conserve these pollinators, which are essential for our food production and agricultural economy.

Winter Firefly,
Ellychnia corrusca

Top: Eastern Comma, *Polygonia comma*

Bottom Left: Green Stink Bug, *Chinavia hilaris*

Bottom Right: Ghost Tiger Beetle *Ellipsoptera lepida*

What is a Native Plant?

◇◇◇◇◇◇◇◇◇◇◇◇◇◇◇

Before the widespread Euro-American settlement in Ontario and across North America, all plants were native. The activities of Indigenous people did influence the region's ecosystems. However, it was in the mid-1800s that large-scale habitat alteration and the introduction of non-native plants began to significantly transform Ontario's natural landscape.

Native plant species in Ontario have evolved over millennia, making them best suited to the region's climate and soil conditions. More importantly, they have co-evolved with native insect species, providing food resources for many insects and invertebrates. These, in turn, serve as food for native birds and other animals. For instance, native plants offer nectar, pollen, foliage, and seeds to feed native insects and other wildlife. Most insects that eat plants can only develop on specific native plant species (host plants). Almost all native birds feed their young insects for protein, regardless of their adult diet. Native bees and other pollinating insects are crucial in providing food for human consumption by pollinating crops such as fruits and vegetables. They also pollinate 90% of all flowering plants, making them indispensable to diverse ecosystems. This intricate web of life, where each species depends on the other, underscores the importance of preserving our native plants and pollinators.

Native plants, with their remarkable adaptability to the climate and soil conditions of their natural habitats, stand as a testament to nature's resilience. They thrive without the need for fertilizers or pesticides and demonstrate a remarkable ability to conserve water, making them a sustainable choice for any landscape.

Native plants have functioning reproductive structures and viable seeds. Non-natives, including nativars and cultivars, are typically bred for specific desired traits. When particular traits are selected, other characteristics can be deselected and disappear from a plant in the nursery trade. This can lead to the loss of genetic diversity in plants available for landscaping or other purposes.

The complex root systems of native plants contribute to healthy soil and reduce and filter water runoff, which protects streams. Choosing native plants for landscaping beautifies yards and other spaces, supports nature's web of life, manages stormwater, and stores carbon. Prairie grasses and wildflowers store carbon in their roots (and in surrounding soil) and, in many situations, do so much more effectively than trees and wooded landscapes.

It is crucial to understand that preserving original habitats is not just a matter of conservation but a necessity for the continued existence of genetically diverse, native seeds for the native plant industry. As gardeners and environmental enthusiasts, we are responsible for encouraging and supporting native landscaping. By doing so, we protect original native plant habitats, including prairie remnants, meadows, wetlands, woodlands, and forests, ensuring their survival for future generations.

Top: Black-eyed Susan, *Rudbeckia hirta*

Middle: Prairie Smoke, *Geum triflorum*

Bottom: Spotted Bee Balm, *Monarda punctata*

Guide to Non-Native Plants

◇◇◇◇◇◇◇◇◇◇◇◇◇◇◇

The natural world is a testament to the awe-inspiring genetic diversity within and among plant species. This diversity, honed over millions of years, directly responds to the unique growing conditions in ecosystems and natural habitats, including geology, soil types, climate, rainfall, and the delicate balance of animal/ insect herbivory and plant diseases.

Native plant species that reproduce in the wild without human intervention play a vital role in our ecosystems. Their genetic diversity equips them with adaptability, making them more resilient in the face of disease, climate change, and other challenges. Moreover, their long-standing co-evolution with insects and other animals ensures the nutritional value of their foliage, pollen, nectar, fruits, and seeds.

Invasive Plants: Invasive plants, a product of global trade, human and animal transport, and gardening, pose a significant threat. These non-native trees, shrubs, and herbaceous plants invade forests, stifling the growth of native plants and disrupting the delicate balance of ecosystems, native vegetation, and native wildlife.

Cultivar: A plant selected for a specific trait, for example, flower colour, foliage colour, fruit colour, shape, size, pest resistance, growth habit, disease resistance, longer bloom times, or stronger stems. Cultivars can be derived from non-native plants or native ones. Cultivars derived from native plants are often called "nativars." Cultivars can have sterile flowers and produce no seeds. Most cultivars are created by cloning (asexual reproduction, such as with plant cuttings), in which clones of the parent plant are made. Some cultivars are strains (seed-grown, relatively true to type) or hybrids via genetic manipulation at the cellular level in a laboratory, such as Roundup-ready corn and soybeans.

Some cultivars, however, are named strains of native plants found in nature and are seed-grown. For example, Eastern redbud (*Cercis canadensis*) 'Columbus strain' is a cultivar valued for its cold hardiness.

Nativar: A cultivar of a native plant, the result of human selection for a specific plant trait(s). Some nativars can have sterile flowers and seeds.

Are cultivars/nativars great or bad for our pollinators and wildlife?

This is a complex question; sadly, there is no straightforward answer.

With cultivars/nativars, humans play a crucial role in selecting certain plant traits over others. The traits selected against (and, therefore, absent in the plant) may or may not have ecological consequences. This underscores the importance of our responsibility in making these choices.

In some cases, cultivars/nativars do not provide food sources for leaf-eating and pollinating insects as their native counterparts do. Tallamy has noted that nativars bred for brown or other dark foliage may not offer the same nutritional value as leaf-eating insects that depend on specific native plants for food. Gardeners should avoid these cultivars if gardening aims to support particular insects that rely on specific host plants to survive.

For example, the 'Summer Wine" ninebark (*Physocarpus opulifolius*) nativar with dark foliage may not be as nutritious as the straight Ninebark, Physocarpus opulifolius species. To give other examples, many nativars have been bred to have more petals and much reduced floral reproductive structures than their native counterparts, with less or perhaps no nectar or pollen for pollinating insects or songbird seeds. Insects may be attracted to plants such as this and use energy travelling to alight on the flowers of this cultivar when, in fact, its flowers may offer little or no food sources. Other nativars are bred for petal colour, which is widely divergent from the colour of the flower in the native species and may confuse the insects and other wildlife that depend on them.

Western Honey Bee,
Apis mellifera

Plant Hardiness Zones

◇◇◇◇◇◇◇◇◇◇◇◇◇◇◇

Recent updates to Canada's plant hardiness zones are a significant development for all gardeners and horticulturists. These updates, incorporating climate data from 1981 to 2010, have introduced zone designations across most areas, ensuring our understanding of plant suitability remains current and relevant. The original hardiness zones map, developed in the 1960s by Agriculture Canada, was based on survival data for various plant species. These zones play a vital role in guiding planting decisions and determining which perennial plant species are suitable for different regions. The United States Department of Agriculture (USDA) has also updated its hardiness zone map. The Canadian and USDA plant hardiness maps have been updated for the Canadian land base using recent climate data and can be accessed at planthardiness.gc.ca.

It's important to note that Canadian and USDA zones do not overlap, meaning that a zone 4a designation using the Canadian approach does not necessarily correspond to a USDA zone 5a. While these maps are useful, they do have limitations. They apply a single formula for the entire country, which may not account for potential variations in climatic variables. Additionally, the hardiness zone designation for a particular plant is often based on factors other than extensive field testing, which can impact the system's overall effectiveness. As the climate changes, testing and assigning plants to zones becomes a dynamic process, highlighting the challenges we face in plant cultivation and wildlife conservation.

To address these limitations, a new approach called 'species-specific distribution models' is emerging as a practical solution. These models use data from a plant hardiness project covering the U.S. and Canada to map potential distributions of specific plant species. This is an encouraging development, as climate envelope models for nearly 3,000 species have been developed. This ongoing work holds the promise of reaching a much larger set of individual species with minimal coordination between nursery growers and citizens, offering hope for the future of plant cultivation.

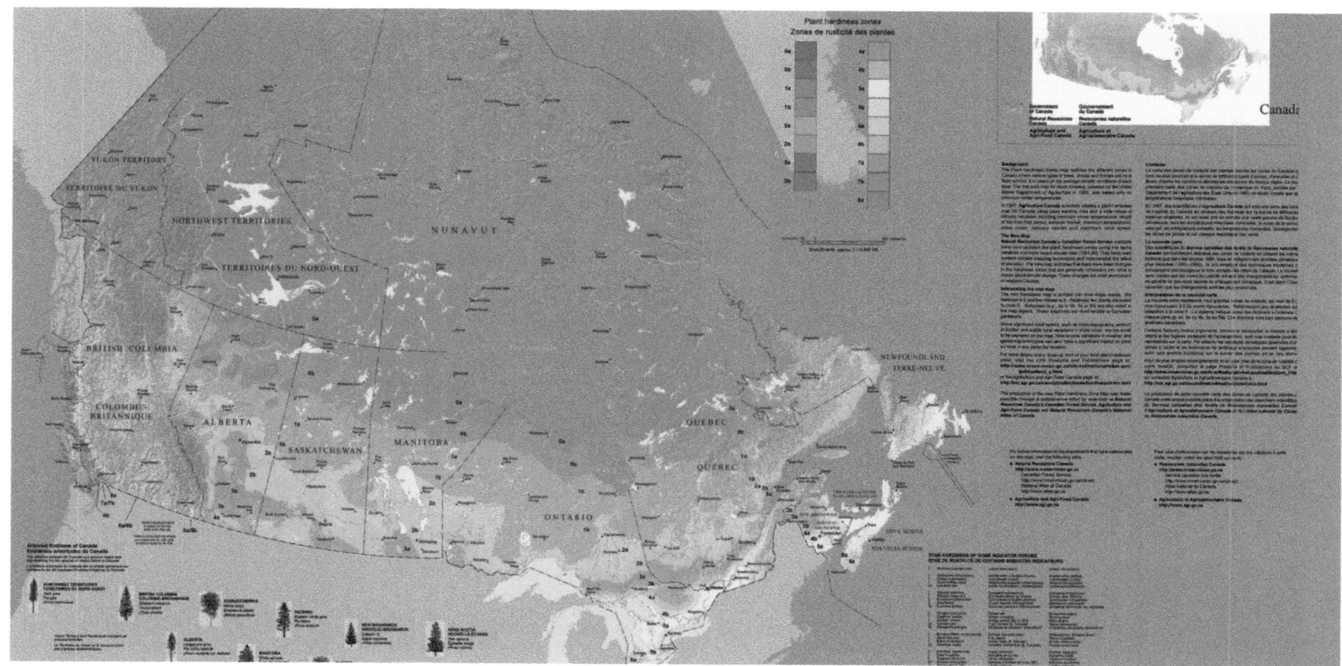

Ontario's Diverse Climatic Conditions

Southern Ontario

The province's southern region has a humid continental climate with distinct seasonal variations influenced by the nearby Great Lakes. During the summer, the area experiences a wide range of temperatures, from warm to hot, often accompanied by high humidity. Winters in this region bring about mild to cold temperatures. Additionally, the area is prone to extreme weather events such as thunderstorms, droughts, and damaging winds, contributing to its dynamic climatic conditions.

Central Ontario

Central Ontario experiences a climate with warm to hot summers and long, cold winters accompanied by heavy snowfall. The arrival of spring brings the risk of river flooding, impacting different regional areas.

Northern Ontario

The northern region of Ontario is renowned for its cold subarctic climate, characterized by short, mild summers and prolonged, harsh winters accompanied by substantial snow accumulation. Additionally, this area witnesses significant fluctuations in temperature across the four seasons.

On average, the province's frost-free period spans from mid-to-late May to late October. This period is crucial for gardening as it signifies the time when the risk of frost is significantly reduced, allowing for the safe planting of frost-sensitive plants. Given the substantial variations in climate across the province, residents and stakeholders must consider the specific frost dates of their particular zone and region.

Challenges of Growing in Ontario

Let's delve into the intricacies of gardening in different regions of Ontario.

Gardening in Southern Ontario can be a real challenge, primarily due to the potential for extreme weather conditions such as droughts, high winds, and thunderstorms. It's essential to monitor the weather closely and consider providing a protective covering for your garden to prevent damage caused by these weather events.

In Central and Eastern Ontario, gardeners face a shorter growing season, heavy snowfall, and spring flooding as the main challenges. To address these issues, raised garden beds can help extend the growing season by providing better drainage and warming the soil faster in the spring. Protective coverings, such as cold frames or row covers, can also be an effective strategy to safeguard your plants from frost and snow and extend the growing season. Despite the subarctic climate and its challenges, gardening in Northern Ontario can be an enriching experience. By maximizing growing time through indoor planting and protective coverings, gardeners can nurture a thriving garden, a beacon of hope in adversity.

Cultivating Conservation

◇◇◇◇◇◇◇◇◇◇◇◇◇◇◇◇

One of the most crucial steps in establishing a garden that supports pollinators and wildlife is carefully selecting a diverse range of native plants. These plants, having evolved naturally in a specific region over thousands of years, are perfectly adapted to the local climate, soil conditions, and wildlife. This adaptation makes them more resilient and less demanding regarding water, fertilizer, and pesticides. Native plants, which have co-evolved with local pollinators, serve as a consistent and varied food source. By opting for native plants, gardeners can actively contribute to preserving local biodiversity, a key element in maintaining a healthy garden ecosystem.

Creating habitat diversity is not just a step; it's a decisive action. Gardeners can achieve a diverse habitat by preserving patches of undisturbed vegetation and installing bee houses or insect hotels. When properly maintained, these provide shelter for a variety of beneficial insects. Reducing pesticide use and opting for organic alternatives is not just a choice; it's a significant contribution. This practice helps maintain the delicate balance of ecosystems, safeguarding pollinators and other wildlife. Water features like birdbaths or small ponds further enhance biodiversity, attracting amphibians, birds, and wildlife. By making these changes to your gardening practices, you can significantly impact the plants you grow and the environment, and that's a reason for hope.

Changing the mindset toward pests is also necessary. Most people think having insects around a garden, even leaving very distinct marks on leaves, is destructive. However, their presence in your garden means you have done well in balancing the forces of nature, even if some plants are damaged in the process. You should welcome them if you desire your backyard to be a sanctuary for wildlife and pollinators. By adopting these multifaceted strategies, such as planting a variety of flowers to attract different pollinators, providing water sources like birdbaths or small ponds, and reducing pesticide use, gardeners can bolster pollinator populations and foster a thriving and harmonious environment for a broad spectrum of wildlife in their gardens.

What Can Gardeners Do?

Plant Native Species: Choose native plants for garden or landscaping projects. Native plants have co-evolved with the local pollinators, providing them with the necessary food sources, such as nectar and pollen. Native plants also support other wildlife, including birds, insects, and mammals, by providing habitat and food. A great way to source some Ontario-specific native plants is from Ontario Native Plants.

Create A Diverse and Layered Garden: Aim for various plants with different bloom times, flower shapes, and colours. This diversity will attract a wide range of pollinators with varying preferences. Include plants that produce flowers throughout the year to ensure a continuous food supply for pollinators.

Provide Habitat Features: Incorporate features in the garden that offer shelter, nesting sites, and overwintering areas for pollinators and other wildlife. Features include planting native shrubs and trees, leaving patches of bare ground for ground-nesting bees, providing deadwood or brush piles, and installing nesting boxes for cavity-nesting bees or birds.

One of the most impactful actions gardeners can take is to minimize or eliminate the use of pesticides in their gardens. Pesticides, including insecticides and herbicides, can harm pollinators and other beneficial insects. By focusing on natural pest control methods, such as attracting beneficial insects that prey on pests like ladybugs or lacewings or using organic alternatives like neem oil or insecticidal soap, gardeners can protect their gardens and the environment without compromising the health of pollinators and other beneficial insects. Remember, these methods are not just alternatives but effective and safe, making them a significant contribution to the overall health of your garden ecosystem.

Provide Water Source: A water feature in the garden, such as a shallow bird bath or a small pond, provides drinking and bathing opportunities for wildlife, including pollinators. Ensure a surface area by offering rocks or floating platforms for easy access for smaller insects, such as bees and butterflies. These water features provide essential hydration for wildlife and serve as attractive elements that can enhance the beauty and biodiversity of your garden.

By actively supporting local conservation efforts, you're not just preserving natural habitats like forests, meadows, wetlands, and wildflower-rich areas. You're also providing crucial resources for a wide range of wildlife, including pollinators, and contributing to the overall health of our environment. As an individual, your actions are invaluable. By sharing knowledge and raising awareness about the importance of pollinators, you can inspire others to adopt pollinator-friendly practices and create wildlife-friendly habitats in their own spaces. This collective effort can make a significant difference in promoting biodiversity and supporting local ecosystems.

By actively supporting local conservation efforts, you're not just preserving natural habitats like forests, meadows, wetlands, and wildflower-rich areas. You're also providing crucial resources for a wide range of wildlife, including pollinators, and contributing to the overall health of our environment. As an individual, your actions are invaluable. By sharing knowledge and raising awareness about the importance of pollinators, you can inspire others to adopt pollinator-friendly practices and create wildlife-friendly habitats in their own spaces. This collective effort can make a significant difference in promoting biodiversity and supporting local ecosystems.

Certify a "Wildlife-friendly garden" with the Canadian Wildlife Federation and get an official sign to help raise awareness in your yard. Consider becoming a "Monarch Waystation" with Monarch Watch to help raise awareness and support for monarchs.

CWF - *Wildlife - friendly Habitat*
https://cwf-fcf.org/en/explore/gardening-for-wildlife/action/get-certified/

Monarch Watch - *Monarch Waystation*
https://www.monarchwatch.org/waystations/

Elevate all gardening actions by incorporating essential strategies that benefit the plants, contribute to a flourishing ecosystem and bring beauty and joy to your surroundings. Opt for native plants that are well-adapted to your region and attract local pollinators, which is crucial for plant reproduction. Diversify the garden's habitat by incorporating features like bird feeders, water sources, and shelter, creating a haven for various wildlife. Imagine the delight of seeing butterflies and birds flocking to your garden, adding a touch of nature's beauty to your everyday life! Other essential strategies include regular watering, proper pruning, and timely pest control, all of which contribute to your garden's overall health and beauty.

Minimize negative impacts by reducing or eliminating pesticide use, opting for natural alternatives whenever possible. Embrace sustainable practices such as composting and mulching. Composting is recycling organic materials, such as kitchen scraps and yard waste, into a rich soil amendment. Conversely, mulching involves covering the soil surface with organic material, such as straw or wood chips, to help retain moisture, suppress weeds, and regulate soil temperature. These practices enrich soil health and promote a thriving garden ecosystem. Remember, every small action contributes to the overall well-being of the garden and the environment!

Right: Red Admiral,
Vanessa atalanta

Building a Wildlife Garden

◇◇◇◇◇◇◇◇◇◇◇◇◇◇◇◇

Regardless of the size of your space, whether it's a small apartment balcony, a yard awaiting landscaping, or several acres of land, your actions can make a significant impact. By following this straightforward, step-by-step guide, you can create your pollinator garden and contribute to the preservation of pollinators for the future. Remember, every effort you make counts, and you have the power to make a difference in the environment around you.

Plan your Garden!

Emphasize the significance of planning in your garden's success. By investing time and effort in meticulous planning, you lay the groundwork for a thriving garden. This step is crucial as it helps you understand your garden's needs and ensures you are fully prepared before planting.

Choosing your location.

When selecting plants for your garden, it's essential to consider the preferences of the intended audience – in this case, butterflies and other pollinators. While flowering plants can thrive in both shady and sunny areas, it's worth noting that butterflies and other pollinators tend to prefer sunny locations. Some of their favourite wildflowers grow best in full or partial sun, with the added benefit of some protection from the wind. Therefore, when planning your garden to attract these beautiful creatures, consider incorporating a variety of flowering plants that flourish in sunny, sheltered spots.

Identifying your soil type and direction of sunlight.

When assessing your soil, it's essential to consider its composition. If it's sandy and well-drained, it will have different implications for plant growth compared to soil that is more clay-like and tends to be wet. You can conduct a simple test by turning over a small patch of soil or utilizing resources like the soil mapper in your county to gather more detailed information. Understanding your soil type and the sunlight it receives is crucial in determining the most suitable plants for your garden or landscape.

Choosing your native plants.

Be sure to conduct thorough research on the specific varieties of milkweed and wildflowers that are native to your area and are well-suited to your soil and sunlight conditions. Native plants are highly recommended because they typically require less maintenance and are more resilient. Look for a reputable nursery specializing in native plants in your vicinity, as they will have extensive knowledge of plants that thrive in your region. Selecting plants that have not been treated with pesticides, insecticides, or neonicotinoids is crucial. Additionally, prioritize choosing perennials to ensure your plants return yearly and demand minimal maintenance.

Consider the entire growing season, not just summer. Pollinators require nectar from early spring to late fall. By selecting plants that bloom at different times, you can create a beneficial garden with a vibrant and diverse display of nature's beauty. This garden will attract and support pollinators throughout the year, providing a constant source of wonder and appreciation.

Which is the best option for you for seeds vs. plants?

Once you have identified the specific plant species you want to grow, the next step is to embark on a journey of patience and dedication. Whether you start from seeds or opt for small plants, each option has unique rewards and challenges. Your decision will not only depend on your timeline and budget but also on the sense of accomplishment you seek from your gardening experience.

Seeds can be a more economical choice, especially if you have a more extensive garden to tend to. However, it's important to note that growing plants from seeds will require more time and patience. If you settle on seeds, it's best to plan for their dispersal in the fall or late winter, well ahead of your anticipated summer growing season. This will allow the seeds ample time to germinate and establish themselves.

On the other hand, nursery-started plants may have a higher initial cost, but they often provide a quicker return on investment. Additionally, these established plants can attract pollinators to your yard during the same growing season, which is an added benefit to consider.

Understanding your unique gardening goals is the key to making an informed decision about seeds or plants. By considering these goals alongside the advantages and considerations of each option, you can make a choice that aligns perfectly with your needs and preferences.

Great Spangled Fritillary,
Argynnis cybele

Planting your Garden

When you're prepared to begin planting, gather your seeds or plants and essential gardening tools, such as a trowel or shovel, to break up the soil. Additionally, you'll want to have extra soil or compost on hand to ensure that your plants have enough nutrients to thrive. Remember to have some mulch available to help retain moisture and keep weeds at bay.

Prep your garden.

Remove the grass and any existing plant cover when transforming a lawn into a garden. Then, turn the soil to loosen it up, creating a welcoming environment for your plants. Numerous pre-made options are available if you're considering raised beds or containers. Alternatively, you can opt for simple designs to build your own. Whichever method you choose, remember to enrich the soil with nutrient-rich compost or quality soil to set the stage for a successful garden.

Planting your seeds or flowers.

When working with seeds, it's essential to consider their germination time. Fall and late winter are the best times to start. In the fall, spread the seeds and cover them with soil. During late winter, scatter the seeds over the snow. The sun will warm the seeds and help them take hold of the snow. As the snow melts, the moisture will aid in seed germination.

If you're using small plants, be mindful of frost dates to prevent planting too early. Dig holes large enough for the root system, then carefully cover and support the roots with soil or compost. Applying mulch can also help minimize weed growth.

Wait, watch, water and weed!

Remember to be patient while waiting for the delightful sight of butterflies and other pollinators enjoying your garden. Watching your garden grow and thrive is a joy in itself. Regularly tending to your garden, weeding, and watering is not just a chore but a way to connect with nature and witness the breathtaking beauty of life.

Right: Great Blue Lobelia,
Lobelia siphilitica

Coleoptera

Exploring Ontario's Beetle Kingdoms

◇◇◇◇◇◇◇◇◇◇◇◇◇◇

As the sun rises, the dew-kissed leaves of Ontario's gardens come alive with the vibrant hues of beetles. With their captivating colours, these insects are a visual delight and essential players in our garden ecosystems. Consider the ladybird beetle, a striking red and black beauty diligently tending to aphid infestations. Or the dung beetle, a diligent worker enriching the soil with essential nutrients. Their presence in our gardens is fascinating and crucial for maintaining a balanced and thriving ecosystem.

Step into the concealed realm of beetles, where the iridescent Emerald ash borer (*Agrilus planipennis*) gleams like polished metal, signalling an alluring yet dangerous presence. Beware these elusive predators as they track pests threatening your cherished crops and other valuable insects. These diminutive creatures are not mere visitors to your garden; they are significant characters intricately woven into the ongoing narratives of your garden.

Delve into the intricate relationship between beetles and flowers. Consider the scarab beetle, a charming insect engaged in the vital pollination dance that sustains life in your garden. Each beetle species represents another captivating chapter in this ever-evolving tale that unfolds with the changing seasons.

Recognize these often overlooked heroes as guardians that contribute substantially to the biodiversity in your backyard haven. In southern Ontario, your garden transcends mere land; it transforms into a stage where beetles assume central roles amid its breathtaking beauty. The diversity of beetles, each playing a unique part in nature's enchanting drama, is a wonder to behold and appreciate. To attract beetles to your garden, consider planting various flowers and providing suitable habitats, such as logs or leaf litter, for them to thrive. These simple steps can make a significant difference in promoting biodiversity in your garden.

Six-spotted Tiger Beetle,
Cicindela sexguttata

Top Left: Goldenrod Leaf Beetle, *Trirhabda canadensis*

Top Right: Goldenrod Soldier Beetle, *Chauliognathus pensylvanicus*

Bottom: Grapevine Beetle, *Pelidnota punctata*

The World of Beetles

◇◇◇◇◇◇◇◇◇◇◇◇◇◇

The world of beetles, belonging to the Order Coleoptera, is a wondrous display of nature's ingenuity and diversity. The name "Coleoptera" finds its origins in the Greek words "koleos," meaning sheath, and "ptera," meaning wings, perfectly encapsulating the essence of these remarkable insects. Beetles are distinguished by their forewings, known as elytra, which serve as protective covers for their delicate hind wings. These elytra act as a sheath, safeguarding the hind wings and body, and provide insight into the origin of the name for this diverse and abundant Order of insects.

Beetles, or Coleoptera, undergo complete metamorphosis, progressing through four distinct stages: egg, larva, pupa, and adult. Eggs are typically laid in or near suitable food sources, and the larvae, known as grubs, undergo multiple moults before pupation. During pupation, the larva transforms into an inactive pupa, from which an adult beetle eventually emerges, marking a remarkable transition from its larval form. This life cycle is a testament to the resilience and adaptability of beetles, and understanding it can help gardeners appreciate the intricate processes at work in their gardens.

In the vast province of Ontario, a sanctuary for beetles, the Order Coleoptera unveils its diversity with unparalleled richness. Over two thousand beetle species have been meticulously identified, each contributing to the ecological tapestry of this Canadian landscape. These beetles thrive in various habitats, from woodlands to meadows to backyard ponds, showcasing their incredible adaptability and resilience, a testament to their survival strategies.

Picture the awe-inspiring iridescence of Tiger beetles (*Family Cicindelidae*) shimmering in the sunlight, their vibrant hues captivating any observer. Equipped with specialized adaptations for swift movement, these agile predators engage in a perpetual dance with the environment, showcasing nature's remarkable efficiency in design. Moreover, consider the cryptic camouflage employed by various beetle species, a testament to the evolutionary arms race between predator and prey, adding intricate layers to the vibrant world of Ontario beetles. The beauty and diversity of beetles are indeed a sight to behold, and by learning more about them, you can deepen your appreciation for the natural world.

As you explore the hidden corners of your garden, be prepared to encounter these living treasures, each with its unique story waiting to be discovered. Ontario's tapestry of beetle life extends beyond mere numbers, representing various forms, behaviours, and ecological roles. However, it's important to note that beetles, like many other insects, face habitat loss and climate change threats. By understanding these challenges, we can take steps to protect and preserve these vital garden inhabitants, ensuring the continuation of their fascinating stories in our gardens.

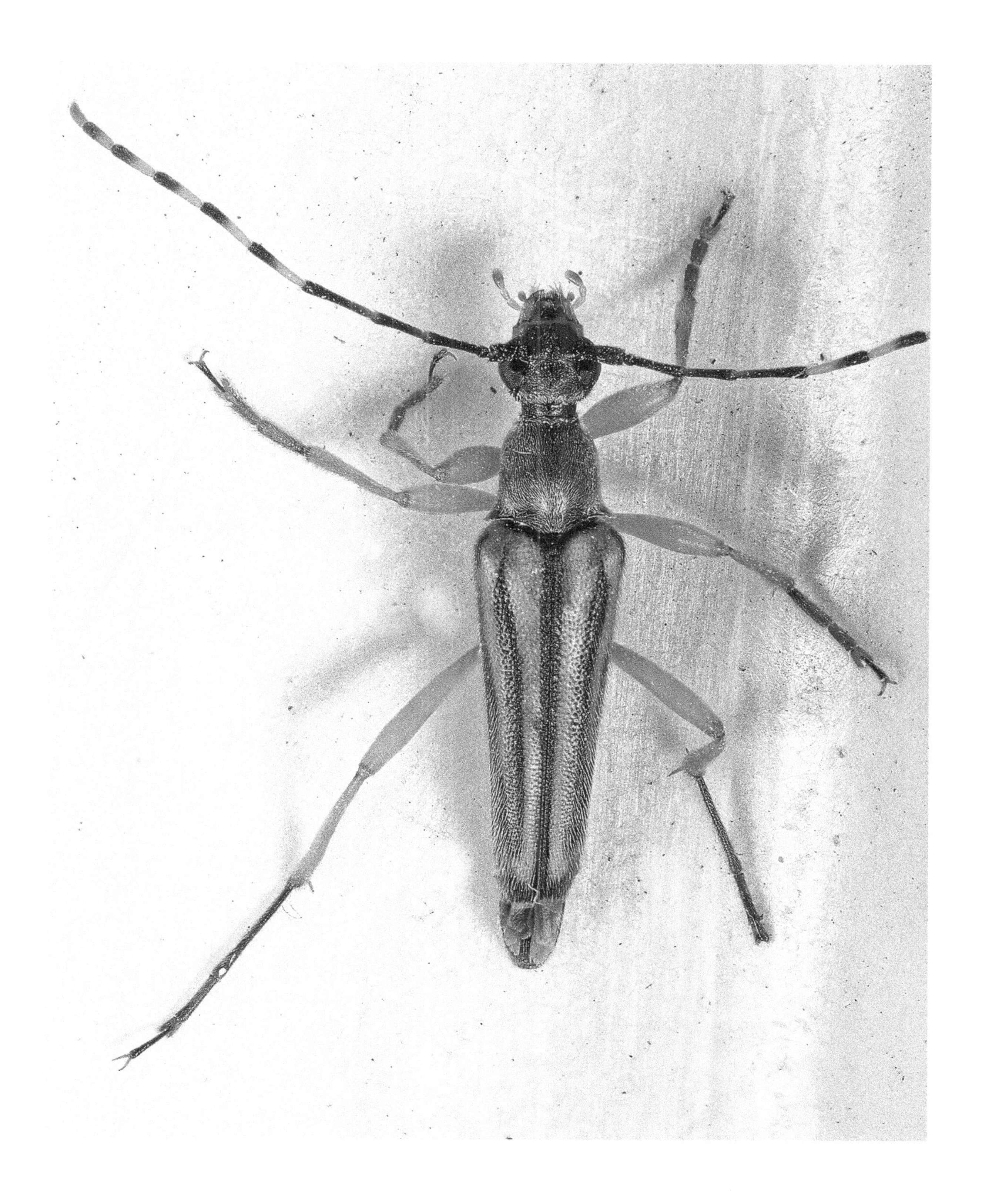

Lined Long-horned Beetle,
Analeptura lineola

Beetles in the Garden

◇◇◇◇◇◇◇◇◇◇◇◇◇◇◇

The multifaceted role of beetles within our gardens and yards extends far beyond their seemingly unassuming appearance. They play a crucial part in maintaining a balanced and thriving ecosystem. One of their paramount contributions lies in natural pest control, where many beetle species emerge as voracious predators, targeting common garden pests such as aphids, caterpillars, and harmful beetle larvae. Their ravenous appetites act as a natural control mechanism, adeptly keeping pest populations in check and reducing the necessity for chemical pesticides. This not only safeguards the garden's health but also minimizes environmental impact.

Today, their taxonomic Order, Coleoptera, makes up the most significant Order in the world. Gardeners have yet to intentionally draw beetles to their gardens, possibly because beetle watching isn't as inspiring as butterfly or bird watching. Yet, beetles do play a role in pollination. Some have a bad reputation because they can leave a mess behind, damaging plant parts as they eat pollen and other parts of the plant. However, their role in natural pest control and soil health must be balanced. By understanding the potential benefits and drawbacks, gardeners can make informed decisions about welcoming beetles into their gardens.

Beyond their role as pest controllers, beetles become integral players in biodiversity enhancement. Their active participation in the complex ecological dance is a testament to their importance. By contributing to the overall richness and diversity of the garden, they foster an environment that is not only visually appealing but also sustainable and conducive to the well-being of all its inhabitants. For instance, some beetles help decompose, breaking down organic matter and enriching the soil. Others, like the locust borer, play a crucial role in pollination, ensuring the production of fruits and seeds. By understanding these ecological roles, gardeners can appreciate the diverse contributions of beetles to their gardens.

In soil health, certain beetle species emerge as unsung heroes in the decomposition process. By breaking down organic matter, they actively contribute to nutrient cycling, enriching the soil and enhancing its fertility. This natural process becomes a boon for the plants within the garden, promoting robust growth and ensuring a healthy, thriving green landscape.

Big Bluestem,
Andropogon gerardi

The pollination prowess of beetles, though often overshadowed by the buzz of bees, takes center stage as another vital role. Some beetle species, like the Locust borer (*Megacyllene robiniae*), emerge as influential pollinators crucial in pollinating various flowering plants. This is particularly significant for reproducing native garden plants, ensuring the production of fruits and seeds. By engaging in these pollination activities, beetles contribute to the cyclical continuation of plant life, shaping the very fabric of the garden ecosystem.

Beetles also act as champions of ecosystem recycling through their role in organic waste management. By feeding on decaying plant material and other organic debris, they actively participate in breaking down and recycling nutrients. This contributes to the overall cleanliness of the garden environment and maintains a nutrient cycle essential for sustained plant health.

In addition to these ecological functions, beetles serve as bioindicators, offering valuable insights into the health of an ecosystem. A thriving beetle population is often indicative of a well-balanced and biodiverse environment. Conversely, a decline in beetle diversity may signal underlying ecological imbalances that warrant attention, making beetles an essential barometer of ecosystem health.

Beyond their ecological significance, beetles provide rich learning opportunities. Observing these fascinating creatures in their natural habitats allows a deeper understanding of biodiversity and ecology. Studying their behaviours, life cycles, and interactions fosters environmental awareness, creating a connection between humans and the intricate world of beetles.

While acknowledging that certain beetles may cause minor, aesthetic damage or pose challenges to specific crops, their benefits to the garden ecosystem underscore the importance of welcoming them into our outdoor spaces. This approach aligns seamlessly with the principles of sustainable and environmentally friendly gardening practices, fostering a harmonious coexistence between humans and the diverse insect life that contributes to the vitality of our gardens.

Short-winged Blister Beetle,
Meloe angusticollis

Morphological Marvels
Anatomy of Coleoptera

◇◇◇◇◇◇◇◇◇◇◇◇◇◇

As previously mentioned, the hard forewings of Coleoptera, known as elytra, act as a shield to the hind wings lying beneath. This adaptation protects them from physical harm and helps regulate water, enabling their survival even in harsh environments where they abound. Elytra also helps in taxonomic identification, with various colours and patterns that make them attractive to watch for lovers of beauty.

A beetle's body beneath the elytra is divided into three parts: head, thorax and abdomen. Beetles' heads have conspicuous mandibles adapted for chewing, cutting and digging depending on the beetle's feeding habits. The thorax, which houses muscles necessary for movement, illustrates these insects' versatility. Aided by their specific lifestyles, beetles show various leg adaptations, such as the swimming legs of aquatic species or digging forelegs of ground beetles.

Looking deeper into the animal's system, we see that Coleoptera's digestive system is a fascinating subject. Various groups of beetles possess specialized mouthparts and digestive enzymes that allow them to feed upon many substances, ranging from plant materials to decomposing matter. This quality makes these creatures important decomposers within garden ecosystems, facilitating nutrient cycles.

Beetle reproductive organs further exhibit enormous biological diversities among these arthropods. While some species have detailed courtship rituals, others have remarkable means by which copulation does take place. Successful gardening, therefore, involves creating conditions that will promote population sustainability through understanding some of these reproductive strategies.

For instance, going deeper into an example of how some Flower beetles reproduce, like those from the Scarabaeidae family, reveals some interesting adaptations. Flower beetles often have particular copulatory organs like modified genitalia or other appendages. To illustrate this point, one can give examples of plants like flowers, which attract insects by having showy colours and structures that aid during copulation. Moreover, when it comes to mating, flower beetles possess some distinctive behaviours that make them more attractive to watch while doing it. For example, one type of beetle performs mating dances; another responds to environmental triggers in its area to start procreation. Spaces within garden landscapes can support various flower beetle life stages, resulting in greater ecosystem diversity and improved ecological health through increased garden biodiversity.

Finally, it appears that nature had let itself go when it came up with Coleopterology, its morphological wonders! From elytra that protect them to the specialized mouthparts and reproductive adaptations, these are a few examples of the beauty of biodiversity, as shown by beetles. Therefore, appreciating the anatomy of these remarkable insects enhances our understanding of the delicate balance within ecosystems as gardeners. By providing an environment for diverse Coleoptera, we build robust gardens that host different species through accommodation, thus creating a symbiotic relationship with these creatures.

Cicindela sexguttata life cycle, egg, larva, pupa, and adult

Positives and Negatives
Welcoming Coleoptera in our Gardens

◇◇◇◇◇◇◇◇◇◇◇◇◇◇◇

Beetles are a very diverse Order of insects under the classification Coleoptera, with over four hundred thousand identified species worldwide. Canada hosts over nine thousand species of beetles that are widely distributed throughout Canada, making Coleoptera the most significant Order of all insects. While it is true that the presence of such insects in gardens could bring about both positive and negative effects, one must understand their role in ensuring a balanced ecosystem. This guide informs gardeners of the benefits and drawbacks of dealing with Coleoptera.

Beetles make a significant contribution to natural pest control. Some species, like ladybugs (*Family Coccinellidae*), ground beetles (*Family Carabidae*) and rove beetles (*Family Staphylinidae*), are ravenous predators that eat garden pests such as aphids, caterpillars and slugs. Introducing these beetles allows pest populations to be controlled without chemical interventions. Furthermore, several wood-boring beetles, such as flower beetles (*Family Scarabaeidae*) and longhorn beetles (*Family Cerambycidae*), also perform a vital role in pollination, which indeed assists various types of flowers to reproduce and improves overall biodiversity within the garden. Another good thing is that concealed beetles, such as dung beetles (*Family Scarabaeidae*) and burying beetles (*Family Silphidae*), help with decomposition. These beetles help decompose various organic matter, thus improving soil structure and fertility.

But there are also possible disadvantages. Certain species of beetles, such as the Colorado potato and flea beetles that feed on leaves, stems, or roots, can cause harm to crops. Monitoring and overseeing the implementation of control measures is a significant step in minimizing potential damage. Woody plants can be attacked and damaged by longhorn and bark beetles (*Subfamily Scolytinae*). Most are harmless, but some species weaken or even kill trees by crowding them out, which creates the need for aggressive strategies. Some beetles, such as blister beetles (*Family Meloidae*), emit chemicals used for their protection that may cause rashes or allergic reactions in some people; handling any unfamiliar type of insect species with care is essential. Furthermore, invasive beetles such as the hardier and stronger multicoloured Asian ladybird (*Harmonia axyridis*) out-compete for the same food source as our native lady beetle species.

Incorporating diverse beetles into your garden can lead to more ecological diversity and offer many advantages, such as killing pests naturally and pollinating. Nevertheless, careful monitoring and effective management structures are essential in identifying potential risks while ensuring a healthy garden environment.

Top: Seven-spotted Lady Beetle, *Coccinella septempunctata.*

Middle: Six-spotted Tiger Beetle, *Cicindela sexguttata.*

Bottom: Black Vine Weevil, *Otiorhynchus sulcatus.*

Discover Coleoptera

Ontario's Rich Beetle Diversity

Ontario, Canada, boasts a rich diversity of over two thousand beetle species, each exhibiting a stunning array of colours, shapes, and fascinating behaviours. These include the iconic ladybugs and the elusive longhorn beetles, among many others. A closer look at your garden will likely reveal many captivating beetle species waiting to be uncovered. Let's delve into some of the prominent types of beetles that can be found in Ontario gardens.

Here Is a Closer Look At The Top Five Coleoptera Speices In Ontario Gardens

Asian Lady Beetle
Harmonia axyridis

◇◇◇◇◇◇◇◇◇◇◇◇◇◇◇

Size: Adult range in size between 5.5-8.5 millimetres.

Description Tips: The adult Asian lady beetle, has a distinctive oval and convex shape. They exhibit various colours, including tan, orange, and red hues. These beetles typically sport several black spots on their wing covers, although some may have indistinct or absent spots. One of the most notable characteristics of Asian lady beetles is the black 'M' shape marking on their thorax, situated just above the wing covers.

Harmonia axyridis, were intentionally introduced to North America in the 1970s as a biological control agent to manage agricultural pests. They were first reported in Canada in 1994. Adult Asian lady beetles can be easily distinguished by their unique features. While there are numerous lady beetles, the Asian lady beetlesare they only species in Ontario that can be identified by a black "M" mark on their thorax, just above the wing covers. This distinct marking sets them apart from other ladybugs.

In addition to their unique appearance, Asian lady beetles are efficient predators, both as adults and in their larval stage. They are commonly found in various crops and landscapes and serve as effective natural pest control agents. These beetles feed on various garden pests, including aphids, tiny caterpillars, scale insects, mealybugs, and mites. Their ability to prey on diverse pests makes them valuable allies in integrated pest management strategies.

Overall, the Asian lady beetle's introduction to North America has proven beneficial in controlling agricultural pests, making them a reliable and effective tool in sustainable pest management.

Common Red Soldier Beetle

Rhagonycha fulva

◇◇◇◇◇◇◇◇◇◇◇◇◇◇◇

Size: Adults range in size between 8–10 millimetres.

Description Tips: The Common red soldier beetle, scientifically called *Rhagonycha fulva*, is a striking insect predominantly orange with a slight iridescent shine. Its elytra, or wing coverings, are a deep black with a distinct patch at the end, and its tarsi, or feet, are also black. This beetle has a narrow, rectangular body and relatively long antennae, with females slightly more prominent. Its appearance is further characterized by its elongated shape and how it moves through its environment.

Rhagonycha fulva are highly beneficial insects that play a vital role in controlling pests in the natural environment. These beetles are particularly valuable in Ontario gardens as they primarily prey on smaller insects, such as aphids, which are known to cause significant damage to plants. Their presence is beneficial and essential for maintaining a healthy garden ecosystem in Ontario. Red soldier beetles are diurnal creatures active during the day and frequently seen around flowers, contributing to pollination.

Their population peaks in the summer and fall, showcasing their effectiveness in pest control. Female *Rhagonycha fulva* beetles lay their eggs in the soil. Once hatched, their predatory and aggressive larvae actively seek out and feed on small invertebrates in leaf litter or loose, moist soil. This behaviour further demonstrates their role in maintaining the balance of the ecosystem, and their life cycle is a testament to their resilience and adaptability.

Red Milkweed Beetle
Popillia japonica

◇◇◇◇◇◇◇◇◇◇◇◇◇◇◇◇◇

Size: Adults range in size between 8- 15 millimetres.

Description Tips: The Red milkweed beetleknown as *Tetraopes tetrophthalmus*, is a unique insect with an elongated reddish-orange body adorned with striking black spots. This beetle also has black legs and long black antennae. One of its most intriguing features is its four eyes, two split by the antennae. This distinct characteristic sets it apart from other insects and makes it a captivating creature to observe in the wild.

The Red milkweed beetle, is a stunningly vibrant insect that plays a crucial role in the natural world. It is often spotted near milkweed plants, particularly the Common milkweed (*Asclepias syriaca*), its primary host plant. These mature beetles sustain themselves by consuming milkweed leaves, developing flower buds, and occasionally the stems. Like the Monarch butterfly (*Danaus Plexippus*), the Red milkweed beetle can absorb the toxic substances in the plant's sap, such as cardiac glycosides or cardenolides, into their bodies. This renders them unattractive to potential predators, and their striking red and black coloration is a warning sign as members of the Longhorn beetle (*Family Cerambycidae*), Red milkweed beetles are distinguished by their long black antennae.

While the Red Milkweed beetle may nibble on your milkweed plants, it plays a role in the garden ecosystem. By feeding on the leaves and flowers, they can help control the spread of some milkweed species. Additionally, they are a food source for some birds that can handle the toxins they've accumulated.

Goldenrod Soldier Beetle
Chauliognathus pensylvanicus

◇◇◇◇◇◇◇◇◇◇◇◇◇◇◇

Size: Adult range in size between 9-12 millimeters.

Description Tips: The Goldenrod soldier beetle is a species of Soldier beetle commonly found in Ontario. It is the region's second most prevalent Soldier beetle species, following the Common Red Soldier Beetle (*Rhagonycha fulva*). These striking beetles feature a long, slender body with a vibrant orange coloration. They are characterized by two prominent brown-black spots on their elytra (wing covers). The head of the Goldenrod soldier beetle is visible with its chewing mouthparts and long, straight antennae.

The Goldenrod soldier beetle is often observed on Goldenrod flowers, which gives rise to its common name. However, they will also visit many other flowers, including yellow composites, Queen Anne's lace (*Daucus carota*), Milkweed species, Rattlesnake master (*Eryngium yuccifolium*), and other late-summer flowering plants, making it a captivating sight for nature enthusiasts.

These beetles are most active from July to September, with their population peaking in August. They can be spotted in various habitats, such as meadows, fields, and gardens, adding to the diversity of the natural environment.
In ecological terms, soldier beetles like the Goldenrod soldier beetles play a vital role. They primarily feed on the pollen and nectar of flowers, contributing to pollination. Furthermore, they help control pest populations by preying on small insects like caterpillars, eggs, and aphids.

One noteworthy aspect of these beetles is their benign nature. They do not harm plants and do not pose a threat to humans or animals, as they neither bite nor sting. Due to their beneficial characteristics and non-disruptive behaviour, they are a valuable and welcomed presence in any garden or natural setting.

Six-Spotted Tiger Beetle
Cicindela sexguttata

Size: Adults range in size between 12-14 millimetres.

Description Tips: The Six-spotted tiger beetles (*Cicindela sexguttata*) are captivating with their iridescent, metallic green and blue exoskeletons. These striking beetles are distinguished by their long, slender legs, large sickle-shaped mandibles, and prominent, bulging eyes. However, what truly sets them apart is the presence of six small, white spots located at the edges of their elytra, the hardened forewings that cover the delicate hindwings, making this feature exclusive to these particular beetles.

The six-spotted tiger beetle is a highly beneficial predatory insect that plays a crucial role in the natural regulation of insect and arthropod populations. These remarkable beetles are easily recognizable due to their distinct body shape and rapid movements, distinguishing them from the slow-moving emerald ash borers they are often mistaken for. The adult *Cicindela sexguttata* and their larvae are voracious consumers of insects and other arthropods in various environments, including forests, gardens, and fields.

Adult tiger beetles are agile hunters known for their remarkable speed and agility. They can chase down their prey at impressive speeds. They can run so fast that their eyes struggle to keep up, temporarily blinding them and causing them to run only short distances at a time. On the other hand, the larvae are ambush predators, lurking in caves in the soil and lunging at their unsuspecting prey.

These fascinating tiger beetles are most active from spring through early summer and are commonly found in sunny areas such as trails. During the winter, adult beetles overwinter in their larval burrows while the larvae remain underground, rarely seen. The long legs of tiger beetles enable them to swiftly dart across the ground, giving them a distinct advantage in capturing their prey.

This chapter has introduced you to the fascinating world of Coleoptera.

These creatures play vital roles in your garden ecosystem, from preying on pests to pollinating flowers. While some beetles can be garden challenges, understanding their life cycles and habitat preferences is key to effective management. Remember, many beetles are beneficial insects, so consider adopting organic and integrated pest management practices to protect them. By creating a diverse and healthy garden, you'll encourage a thriving population of beneficial beetles while minimizing the need for harmful chemicals.

Planting for Coleoptera

Discover Native Ontario Plants

To create an irresistible haven for beetles in your garden, strategically planting native Ontario species goes beyond mere horticulture—it's an invitation to witness a thriving, interconnected ecosystem. Here's a closer look at some native Ontario plants from Ontario Native Plants that we suggest you, as gardeners, plant and their role in enticing and supporting the diverse world of beetles and other wildlife.

By strategically integrating these native Ontario plants and others into your garden, you're not just adding to its visual appeal but crafting a vigorous stage where beetles play pivotal roles. This carefully curated selection becomes an essential component of your garden's biodiversity, supporting beneficial beetle species and enhancing the overall ecological resilience of your outdoor space. As these plants bloom and flourish, they beckon beetles, turning your garden into a living tapestry where each species contributes to the intricate dance of life, from the ground-dwelling protectors to the aerial pollinators. Your actions are crucial in this delicate balance.

White Wood Aster
Eurybia divaricata

◇◇◇◇◇◇◇◇◇◇◇◇◇◇◇

Growing Habits:

Mature Height: 2 feet
Mature Spread: 3 feet

Ontario Hardiness Zone: 3 to 7

Full Shade to Part Shade

Dry to Medium

Sand Loam, Loam, Clay Loam

The White Wood Aster is a perennial plant with deeply and irregularly serrated leaves. The lower leaves are heart-shaped, while the upper leaves are elongated. It flowers in the fall, typically in early to mid-September. The florets are yellow and purple, while the rays are white.

This species, faces possible endangerment in the Canadian Province of Ontario and is naturally confined to limited locations. Currently, *Eurybia divaricata* is "Threatened," which means this species lives in the wild in Ontario but is not endangered. It is likely to become endangered if steps are not taken to address the factors that threaten it. The White Wood Aster typically thrives in dry deciduous forests where it can be found in association with Sugar Maple (*Acer saccharum*) and American Beech (*Fagus grandifolia*) trees. It spreads through rhizomes and seeds and is well-suited for providing ground cover over expansive areas, especially in dry, shaded areas. One of its most impressive traits is its ability to flourish in shallow rocky soil, showcasing its resilience and adaptability to various environments.

While not exclusively, *Eurybia divaricata* plays a significant role in the ecosystem. It is a beneficial plant for Coleoptera, attracting pollinators, including some beetle species. These beetles find nourishment from the flower's pollen and nectar, contributing to the plant's pollination process. This underscores the importance of conserving the White Wood Aster, as its loss could disrupt the ecosystem.

Canadian Serviceberry
Amelanchier canadensis

◇◇◇◇◇◇◇◇◇◇◇◇◇◇◇

Growing Habits:

Mature Height: 30 feet

Ontario Hardiness Zone: 4 to 7

Part Shade to Full Sun

Medium

Sand Loam, Loam, Clay Loam

Amelanchier canadensis, a dense, upright, multi-stemmed large shrub or small tree, presents a delicate, dome-shaped crown that is a sight to behold. In mid-spring, it adorns itself with attractive erect sprays of small, slightly fragrant, white, star-shaped flowers, a prelude to the emergence of its leaves. These flowers, a magnet for pollinators, are succeeded by small, edible, blue-black berries in early summer. These berries, a culinary delight in jams, jellies, and pies, are equally cherished by birds and humans.

Amelanchier canadensis is more than just a pretty sight; it serves as a beacon for various bird species. The American Robin (*Turdus migratorius*), Baltimore Oriole (*Icterus galbula*), Cedar Waxwing (*Bombycilla cedrorum*), and multiple Warblers (*suborder Passeri*) flock to these trees. With their nutritious berries and welcoming branches, Amelanchier canadensis provides vital sustenance and habitat, making them a crucial part of the ecosystem. We can play a significant role in preserving biodiversity by appreciating and conserving these trees.

In the fall, the foliage of *Amelanchier canadensis*, with its mid-green, finely toothed, oval leaves, transforms into a breathtaking palette of brilliant shades-yellow, orange, and red. The plant remains attractive even after the leaves drop, thanks to its elegant growth habit and light gray bark adorned with charcoal-gray striations. This four-season interest makes it a versatile and exciting choice for use around ponds, lakes, streams, or in boggy or marshy ground.

Amelanchier canadensis is a haven for various Coleoptera throughout the growing season. Its springtime blossoms provide a welcome source of pollen and nectar for beetles emerging from dormancy. The leaves offer sustenance throughout the summer, while the late summer and fall bounty of purplish-black berries become a delicious feast for many beetle species. Amelanchier canadensis is a valuable food resource for a diverse range of beetles.

Big Bluestem
Andropogon gerardii

◇◇◇◇◇◇◇◇◇◇◇◇◇◇

Growing Habits:

Mature Height: 7 feet
Mature Spread: 2 feet

Ontario Hardiness Zone: 3 to 7

Part Shade to Full Sun

Dry to Medium

Sand Loam

Explore Big Bluestem, a warm-season grass indigenous to the eastern United States and Eastern Canada. Thriving from the mid-western short grass prairies to the coastal plain, it plays a natural role in periodic fires. Standing tall and robust, mature plants commonly reach six to eight feet tall. The rhizomes are short and scaly, while the leaves exhibit a spectrum from light yellow-green to burgundy. Notably, the seed head is coarse, distinguishing it from other bluestems.

Andropogon gerardii, known for its use in erosion control plantings, may take a while to establish, but once rooted, it offers outstanding stability in sandy areas. This species is an excellent native choice for grazing forage and is highly palatable to livestock. Like its bluestem counterparts, Andropogon gerardii creates a habitat for various wildlife species, especially ground-nesting birds that find shelter and forage cover in this clump-forming grass.

Andropogon gerardii is a vital habitat for many beetle species, while also being a crucial player in the ecosystem. The tall grasses provide a safe refuge for beetles, shielding them from predators and harsh weather conditions. Moreover, Big bluestem acts as a food source for certain beetle species. The plant's seeds are nutritious for ground-dwelling beetles, while the leaves nourish other beetle varieties. The role of Big bluestem in providing shelter and sustenance is not just crucial but urgent, significantly contributing to the health and diversity of Coleoptera populations.

Bunchberry
Cornus canadensis

◇◇◇◇◇◇◇◇◇◇◇◇◇◇

Growing Habits:

Mature Height: 1 foot

Ontario Hardiness Zone: 2 to 7

Part Shade to Full Sun

Medium to Wet

Organic

Bunchberry, a diminutive perennial flowering plant of the dogwood Family, Cornaceae, is a lifeline for wildlife and a cornerstone of the forest ecosystem. Native to North America, it thrives in cool, moist, shaded forest understories, its low-growing habit forming dense mats of dark green, oval-shaped leaves. Its most distinctive feature is its unique flower structure, a single, white flower that is actually a cluster of tiny, greenish flowers surrounded by four large, white petal-like bracts. This specialized adaptation is a crucial part of the forest ecosystem, attracting larger pollinators, such as beetles. After pollination, Bunchberry produces clusters of bright red berries, a vital food source for various bird species. Despite its small size, Bunchberry is a key player in the forest ecosystem, providing food and habitat for wildlife, and fostering a deep connection and empathy towards nature.

Cornus canadensis is particularly attractive to beetles due to its unique floral structure. Unlike many other flowers that rely on wind or smaller insects for pollination, Bunchberry has evolved a mechanism to attract more giant, heavier insects, such as beetles. The "flower" of the Bunchberry is a cluster of small, greenish flowers surrounded by four large, white petal-like bracts. This structure serves as a visual lure for insects. When a beetle lands on the flower, its weight triggers a rapid opening of the petals, exposing the pollen-bearing stamens. This sudden movement, often called the "bunchberry pop," is reliable for transferring pollen between flowers.

While the primary benefit to the plant is efficient pollination, the flower also provides a source of nectar for visiting insects, including beetles. This food reward further incentivizes beetles to return to Bunchberry flowers. Therefore, combining a reliable food source and the unique pollination mechanism makes Bunchberry a particularly appealing plant for beetle populations.

White Meadowsweet
Spiraea alba

◇◇◇◇◇◇◇◇◇◇◇◇◇◇◇

Growing Habits:

Mature Height: 4 feet

Ontario Hardiness Zone: 3 to 7

Part Shade to Full Sun

Medium to Wet

Loam, Organic

Spiraea alba, commonly known as White meadowsweet, is an upright deciduous shrub reaching about 4 feet. It is characterized by its narrow, serrated green leaves and bears cone-shaped clusters of small white flowers, which typically blossom in the summer. Following the flowering period, the fruits of *Spiraea alba* ripen in September. Each fruit contains five pod-shaped follicles that split open to disperse the seeds. While the fruits are not particularly showy, the plant is highly valued for its ecological significance. Spiraea alba is commonly found in its natural habitat, thriving in wet prairies, wet river bottom prairies, and open areas along streams or lakes.

Spiraea alba provides a welcome food source for some Coleoptera species. These beetles are not just attracted to the nectar and pollen of the nodding onion's flowers, but they also play a significant role in the plant's pollination process, a contribution we should all respect.

By attracting various beetles, White meadowsweet also indirectly benefits the garden ecosystem. These beetles can act as pollinators for other plants in your garden, helping with fruit and seed sets. Some beetles attracted to the meadowsweet might be herbivores feeding on the leaves. However, these herbivorous beetles can become a food source for predatory beetles, ladybugs, and even ground beetles, which are beneficial for pest control in your garden.

By incorporating these native plants into your Ontario garden, you're creating a haven for a diverse range of beetles. Remember, it's not just about attracting beetles, but supporting the entire ecosystem. These plants provide food, shelter, and breeding sites for countless beneficial insects, including beetles. A biodiverse garden is a resilient garden, better equipped to withstand pests and diseases. As you observe your garden, you'll likely discover new beetle species and appreciate the intricate web of life supported by these remarkable plants.

Diptera

Ontario's Diptera Delight

◇◇◇◇◇◇◇◇◇◇◇◇◇◇

The exquisite gardens of Ontario offer a mesmerizing spectacle. It is a realm where countless flies, known as Diptera, dance and flit around the blossoms. These creatures are not mere nuisances but integral components of a finely tuned ecosystem. Before swatting them away, let's delve into the intricacies of their roles as they unveil the very essence of our gardens.

Within this enchanting domain, each petal serves as a stage and every leaf as a canvas on which life unfolds. Diptera, including the remarkable Tiger bee fly (*Xenox tigrinus*), with its delicate wings and graceful movements, plays leading roles in this captivating natural drama. It's in the details often overlooked that the true magic unfolds.

These buzzing insects are not just passing visitors in our gardens; they are intentional contributors to the symphony of nature. Their presence amidst the foliage speaks volumes about the interconnectedness of the ecosystem. Picture a flower gently swaying while its delicate folds conceal Diptera diligently at work: perpetuating pollination, discreetly aiding in decomposition, and delicately managing the food chain. Therefore, before hastily dismissing these winged creatures, let's immerse ourselves in the wonders they bring about. Envision your garden as a living, breathing organism, with each Diptera contributing to its vitality. Whispers of concealed interactions rustle through the leaves of this microcosm, and collaboration fills the air with its rich fragrance.

These gardens have evolved into more than just landscapes; they are home to over 1,500 species of Dipterans, each weaving tales of biodiversity and vibrancy. When we uncover this hidden realm through our senses, even these inconspicuous flies transform into nature's deeply designed ambassadors. So, let yourself stroll through your garden and listen to the resounding hum that will accompany you throughout the day and night. As a result, you will discover the enchantment beneath each petal or leaf – a world teeming with life where Diptera takes center stage in nature's captivating dance.

Dusky-winged Hover Fly,
Ocyptamus fuscipennis.

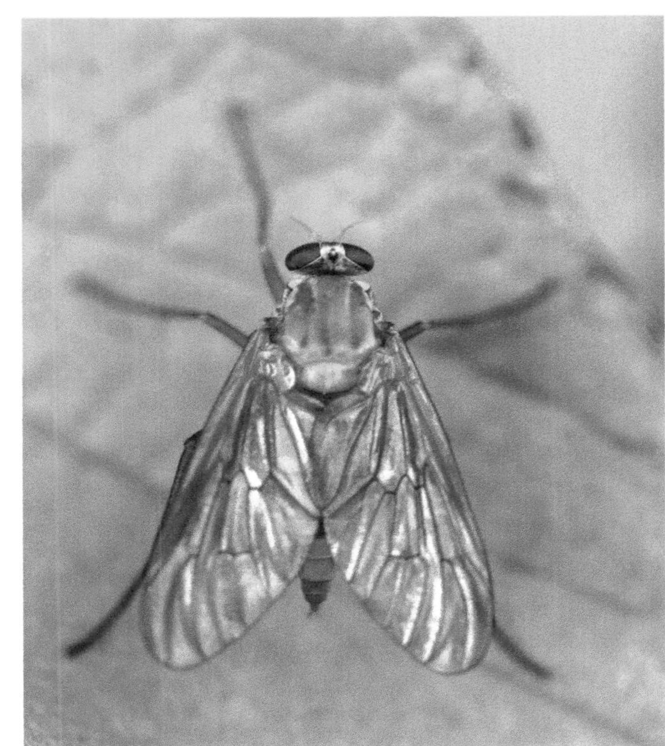

Top Left: Northeastern Hammertail, *Efferia aestuans*

Top Right: Marsh Snipe Fly, *Rhagio tringarius*

Bottom: Common Drone Fly, *Eristalis tenax*

Understanding Diptera

◇◇◇◇◇◇◇◇◇◇◇◇◇◇

The Order Diptera is a fascinating group of insects whose name derives from the Greek words' di' for two and 'ptera' for wings. These insects are a sight to behold, with their distinctive wing structure, where the hindwings have evolved into small club-shaped structures known as halteres. With over 125,000 identified species worldwide, Diptera is a testament to the diversity and adaptability of the insect world.

Diptera, including flies and mosquitoes, undergo complete metamorphosis consisting of four stages: egg, larva (maggot), pupa, and adult. The eggs are laid in various habitats, and the larvae develop in diverse environments. Pupation leads to the formation of a pupa, from which the adult fly emerges, displaying a significant morphological change from the larval stage.

In the entomologically rich province of Ontario, a myriad of Diptera species thrives, each showcasing a unique beauty and contributing to this order's remarkable adaptability and resilience. From the Common housefly (*Musca domestica*) to the elegant and delicate overflies (*Family Syrphidae*) engaging in intricate aerial dances, each species adds a unique hue to the vibrant tapestry of garden life.

For instance, the Common housefly (*Musca domestica*) is often overlooked but plays an intricate ecological role. Belonging to the Muscidae Family, houseflies are prolific decomposers, aiding in the breakdown of organic matter and nutrient cycling, essential to the garden's natural recycling system.

On the other hand, enchanting hoverflies, which resemble stinging bees and wasps, fulfill a crucial role in pollination. Despite being harmless, their graceful flight patterns and nectar-sipping activities make them vital contributors to the reproductive success of various flowering plants. They serve as nature's gentle pollinators and ensure our gardens' continued abundance of blossoms.

Why Do We Need Diptera?

◇◇◇◇◇◇◇◇◇◇◇◇◇◇

The presence of Diptera, or the true flies, in our gardens is not a matter of chance but a vital cog that helps keep delicate ecosystems ticking over. These diligent bugs are multifunctional about our garden's well-being. The primary reason we need Diptera is their critical involvement in pollination. While bees receive most of the attention in pollination talk, many other flies actively participate in this crucial process, especially syrphids or hoverflies. Hoverflies, for instance, are excellent pollinators; they seek nectar from various flowers, and as they do so, they transfer pollen from one flower to another. Their effectiveness in pollination guarantees reproductive success for flowering plants, encouraging biodiversity and supporting the general ecosystem.

Moreover, Diptera is a fascinating part of our gardens and a priceless friend in biological pest control. Some predatory flies, like robbers and predatory hoverflies, gobble up garden pests like aphids and caterpillars, among other harmful insects. Their large appetite makes them efficient at pest control, thus empowering gardeners to manage pests without using chemicals. By encouraging these predator flies into our gardens, we can achieve sustainable and eco-friendly pest management practices that minimize reliance on harmful pesticides, giving us the power to protect our gardens and the environment.

In addition, Diptera's efforts mainly rely on the decomposition process within our gardens. Carrion-feeding flies such as blowflies play an essential role in the breakdown of organic matter, accelerating the decay of animal carcasses and other decaying materials. This aids nutrient recycling because it speeds up the disposal of dead organisms, preventing diseases from infecting others. Also, some fly species, like Black soldier fly larvae (*Hermetia illucens*), help decompose plant matter by breaking down fallen leaves as perishable organic materials, improving soil health.

Diptera's less known yet significant role in our gardens is nutrient cycling. Certain fly larvae, especially crane fly larvae belonging to the Tipulidae family, act as detritivores as they consume soil organic matter, breaking it down and releasing essential nutrients. This intricate process of nutrient cycling is crucial for the health of our plants, as it maintains a balanced and fertile environment suitable for plant growth, a fact that we, as gardeners, should be aware of.

The presence of Diptera in our gardens is not a coincidence but a testament to their vital role in maintaining sound ecosystems. From pollination to biological pest control, decomposition, and nutrient cycling, true flies are integral to our gardens' design. Recognizing and encouraging Diptera in our garden environments is not just a sustainable strategy for ecosystem governance but also a realization of their intricate functions that form the lifeblood of our natural surroundings.

Transverse-banded Flower Fly,
Eristalis transversa

Morphological Marvels
Anatomy of Diptera

Diptera, a unique insect with only two wings, stands out in the insect world. Unlike the usual four, these wings undergo a fascinating transformation, turning the hindwings into small club-shaped structures called halteres. This adaptation defies aerodynamics and provides Diptera with exceptional flight stability, essential for their movement across diverse garden landscapes.

The head of a dipteran insect is a sensory powerhouse, housing specialized structures and organs. The most striking feature is the large compound eyes, which grant a 360-degree view of the surroundings. The antennae, often underestimated, are instrumental in sensing environmental cues. Equally important are Diptera's mouthparts, uniquely adapted for various nutritional strategies. For example, mosquitoes have piercing-sucking mouthparts, while houseflies have sponging mouthparts. These adaptations are a testament to the diverse ecological roles of Diptera.

Diptera, with their captivating feeding preferences ranging from predation to nectar feeding, play a vital role in the garden ecosystem. Their proboscis, accompanying their mouth parts, feeds on various food sources, establishing them as crucial pollinators, decomposers, and predators. Understanding these diverse feeding strategies is vital to supporting biodiversity and health within our gardens.

Moreover, there are interesting reproductive strategies among Diptera species as well. Many species perform elaborate courtship dances employing visual clues and pheromones to attract mates. Diptera's' larval or maggot stages exhibit the ability to tolerate many habitats, enabling them to expand into various ecological niches.

Dipteran anatomy is an example of perfect ecological versatility and evolutionary adaptation. From the unique wings and sensory organs to their various feeding strategies and reproductive behaviours, true flies epitomize the delicate balance of nature. Understanding the morphology of Diptera also helps us appreciate the complexities of ecological systems as gardeners. By creating conditions encouraging these insects to thrive, we support biodiversity and health within our gardens, acknowledging Diptera as integral parts of nature's patchwork quilt.

Common House Fly,
Musca domestica

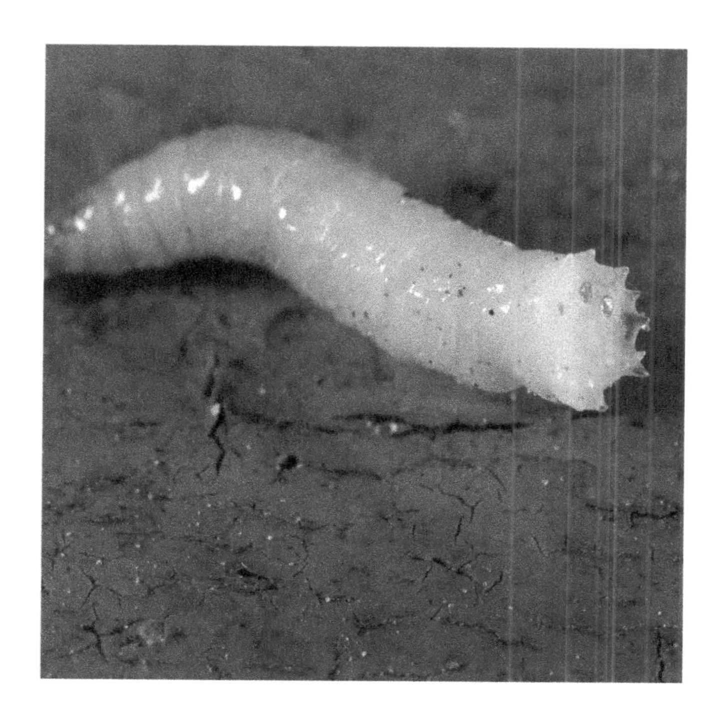

Musca domestica life cycle, eggs, larva, cocoon, adult.

Positives and Negatives
Welcoming Diptera in our Gardens

◇◇◇◇◇◇◇◇◇◇◇◇◇◇◇

Approximately 2,550 species belong to the Diptera Order in Canada and over 1,700 species in Ontario, each performing distinct roles in garden ecosystems. Recognizing the beneficial and detrimental aspects of various species is essential when you have Diptera in your garden.

One of the positive sides is that Diptera species pollinate plants effectively. Though less recognized than bees, some fly species, such as Hoverflies (*Family Syrphidae*) and Bee flies (*Family Bombyliidae*), help transfer pollen between flowers to promote the reproduction of various plants. Furthermore, flies are great decomposers, with several species, such as Blowflies (*Family Calliphoridae*) and Flesh flies (*Family Sarcophagidae*), playing a crucial role in the decomposition of organic matter. This decomposition process is not only helpful in terms of recycling nutrients within the soil but also supports general health.

Nevertheless, Diptera may have some adverse effects on your garden. Certain fly varieties like Fruit flies (*Family Tephritidae*) and Leaf miners (*Family Agromyzidae*) can ruin crops. Fruit flies are attracted to ripe bananas, apples, and other fermented fruits and vegetables; however, they do not consume the fruit. Instead, fruit flies feed on the microorganisms growing outside rotting fruits and other decaying materials. While leaf miners burrow through plant leaves, causing damage that may affect plant health. It is necessary to monitor and control these potential issues properly. Further, some fly species, like the Stable flies (*Family Muscidae*), are a real annoyance as they feed on blood and can transmit diseases to humans and livestock. Proper waste management and sanitation practices can mitigate such concerns.

Incorporating the vast population of Diptera within your garden holds so much potential, from pollination to complex decomposition. However, gardeners must still be alert to possible difficulties, especially regarding the destruction of plants or pest flies. Armed with the knowledge of various roles played by different fly species and adopting appropriate management strategies, gardeners can develop a healthy ecosystem in their gardens.

Orange-legged Drone Fly,
Eristalis flavipes

Middle: Eastern Calligrapher,
Toxomerus geminatus

Botton: New York Bee Killer,
Diogmites basalis

Discover Diptera

Ontario's Rich Diptera Diversity

Ontario boasts over one thousand seven hundred Diptera species, from the prolific Tabanidae Family, commonly known as horseflies, to the delicate Chironomidae, or midges; our gardens host various winged wonders. Here are some of the common Diptera across Ontario gardens.

Here Is a Closer Look At The Top Five Diptera Speices In Ontario Gardens

Eastern Calligrapher

Toxomerus geminatus

◇◇◇◇◇◇◇◇◇◇◇◇◇◇◇◇

Size: Adults range in size between 6.1 to 7.6 millimetres

Description Tips: The coloration of the insect is a remarkable example of Batesian mimicry, a form of protective mimicry where a harmless species evolves to imitate the warning signals of a harmful species. In this case, the flower fly resembles the warning coloration of stinging bees and wasps. The back of the head is predominantly white, with two large compound eyes positioned on each side of the head, along with three small, simple eyes (ocelli) arranged in a triangular pattern at the apex of the head, between the compound eyes. The compound eyes exhibit a reddish-brown hue and are distinctly indented on the rear margin, forming a broad triangular shape. Notably, on the male insect, the compound eyes meet near the top of the face, while on the female, they do not.

The male's face is a striking feature, characterized by a pristine white coloration. In contrast, the female's face is marked by a dark, broad, vertical stripe on the white, extending from between the eyes to the antennae. The antennae, short and yellow, are another distinctive trait. Moving to the thorax, a shield-shaped exoskeletal plate covers the expansive middle section, known as the mesonotum, which is the dorsal part of the thorax. The sizeable anterior portion of this shield, called the scutum, displays a blackish-brown hue with faint, yellowish-brown longitudinal stripes while being notably hairless. Additionally, a small yellow spot, the supracoxal marking, is a common feature in most members of the genus Toxomerus, located on the side of the thorax just above the front leg's first segment, also known as the coxa. The term "supracoxal" denotes its position above the coxa.

Regarding habitat and behaviour, flower flies are a fascinating species. Unlike other fly species, they pose no threat to humans or other animals, as they primarily feed on nectar and pollen from various flowers during their adult stage. These intriguing insects show a preference for small white and yellow flowers. However, the dietary choices of flower fly larvae, often known as aphid lions, exhibit significant variation between species. With their voracious appetite for aphids, these larvae play a pivotal role in natural pest control, underscoring their ecological importance.

Margined Calligrapher
Toxomerus marginatus

◇◇◇◇◇◇◇◇◇◇◇◇◇◇

Size: Adults range in size between 5 – 6 millimetres

Description Tips: The Margined Calligrapher fly, scientifically known as *Toxomerus marginatus,* is a captivating hoverfly species in North America. What sets this species apart is its remarkable adaptability in appearance. Adult individuals take on a darker hue when exposed to colder temperatures, a unique trait among hoverflies. Another distinguishing feature is the yellow abdominal margin, contrasting with the typical black striping extending to the Eastern calligrapher margin (*Toxomerus geminatus*). This small-sized fly also sports a striking black and yellow coloration, lacks a spot near the tip of the abdomen, and boasts an entirely yellow "margin" encircling the abdomen.

The Margined Calligrapher fly and the Narrow-headed Marsh Fly are not just unique in their appearance, but they also play a crucial role in the ecosystem. Their larvae are beneficial predators, effectively controlling various pests such as aphids, mealybugs, thrips, and caterpillars. This natural pest control behaviour makes them invaluable in agricultural and garden settings. As adults, these flies have a diverse diet, feeding on the nectar of a wide range of flowers. This sustains their population and contributes significantly to the vital pollination process, highlighting their ecological importance.

The Margined Calligrapher fly is not just a fascinating species but also a practical solution to pest control. It stands out as one of the most abundant Syrphidae species involved in regulating aphid populations in lettuce fields. Its presence and feeding habits make it a key player in integrated pest management strategies, providing natural and sustainable pest control in agricultural environments. This real-world application of the Margined Calligrapher fly underscores its importance and relevance in pest control, making it a species of significant practical value.

Narrow-headed Marsh Fly

Helophilus fasciatus

◇◇◇◇◇◇◇◇◇◇◇◇◇◇

Size: Adults range in size between 12-15 millimetres

Description Tips: The Narrow-headed Marsh Fly (*Helophilus fasciatus*) is a hoverfly species with unique physical features. The thorax of this species is adorned with prominent black and white lengthwise stripes, creating a striking appearance. Moving on to the abdomen, we see it is decorated with transverse black bands and delicate, lemon-yellow stripes. Notably, the first yellow stripe is always incomplete, adding to the insect's unique patterning. Additionally, while the subsequent yellow stripes are often complete, there is a degree of variability in their completeness, particularly in different individuals. Like other hoverflies, this species undergoes a complex mating and reproductive process involving courtship displays and egg-laying on suitable substrates.

One of the most intriguing features of these flies is their captivating dark red eyes. It's fascinating to note that even in males, the eyes never meet in the middle, setting them apart from other fly species. This unique eye pattern adds to the allure of this particular hoverfly, sparking curiosity about its evolutionary significance.

One of the most reassuring qualities of flower flies, which belong to Syrphidae, is their benign nature. Unlike some other fly species, they do not pose a threat by biting humans or other animals. Instead, these flies primarily feed on nectar and pollen from a diverse array of flowers during their adult stage. Their preference leans towards small white and yellow flowers, showcasing their specific foraging habits.

While the adult diet is relatively uniform across the species, the dietary habits of flower fly larvae display significant variation. As larvae, these flies exhibit a fascinating array of feeding behaviours, reflecting the complexity of their life cycle. Notably, larvae of this species function as aquatic filter feeders, sustaining themselves by consuming decaying vegetation and other organic material in their aquatic environments. This dual-phase feeding strategy highlights the fascinating ecological role played by *Helophilus fasciatus* in its various life stages.

Tiger Bee Fly
Xenox tigrinus

◇◇◇◇◇◇◇◇◇◇◇◇◇◇

Size: Adults range in size between 11 – 19 millimetres

Description Tips: When viewed from the front, the Tiger bee fly's long proboscis makes the insect appear like a giant mosquito. Its body is black and fuzzy, suggesting it is an oversized bee. The fly's long, tapered, swept-back, transparent wings are decorated with intricate black patterns. The markings along the wing's trailing edge, resembling tiger stripes, are a unique feature of this species. Its vast eyes appear to be giant versions of housefly eyes. The insect's black abdomen is marked with two white spots. Despite its fearsome appearance, the tiger bee fly is a fascinating creature worth learning about.

The Tiger bee fly, despite its name, is not a bee but an actual fly. Its menacing appearance and name might lead you to believe this is one critter you must avoid. It is one of the most enormous flies you have probably ever seen. The tiger bee fly's life cycle begins with the female hunting for carpenter bee nest sites after mating. This is why they are often seen hovering near carpenter bee nests. Once the female tiger bee fly finds a nest containing carpenter bee eggs, she lays her eggs alongside the eggs left by the bees. This unique behaviour is a survival strategy for the tiger bee fly's offspring, as they will hatch and parasitize the carpenter bee larvae.

The fly's larvae are parasites. On hatching, they are highly active and seek carpenter bee larvae. The tiger bee fly larvae then attach themselves to newly hatched carpenter bee larvae and consume them. When each tiger bee fly completes this feeding task, all that is left of the carpenter bee is an empty hull. The carpenter bee is a food source for the tiger bee fly's larvae. Although this feeding behaviour might seem intense, it ensures that the carpenter bee larvae never live long enough to leave their nest. Its wing markings are distinctive. Most people see this species hovering around wooden privacy fences, roof overhangs, and similar wooden surfaces.

It is a member of a suite of native pollinators that inhabit our backyards. *Xenox tigrinus* uses its fearsome beak to sip nectar from flowers. If you are looking for an ally to help control carpenter bee damage to your home and property, it makes sense to encourage tiger bee flies to take up residence in your backyard. One of the simplest ways to do this is to stock your yard with various native nectar-bearing plants that provide blooms throughout as much of the year as possible. Also, reduce or eliminate the use of insecticides.

Common Greenbottle Fly
Lucilia sericata

◇◇◇◇◇◇◇◇◇◇◇◇◇◇◇

Size: Adults range in size between 8-10 millimetres

Description Tips: Adult Green bottle flies, with their metallic green or copper green coloration, are a unique sight. Their yellow mouthparts and hairy backs add to their distinct appearance. These flies have hairless squamae at the base of their wings.

The larvae of *Lucilia sericata*, also known as the medicinal maggot, have a wide range of applications. They have been used in managing chronic, non-healing wounds (larval therapy), aiding in forensic investigations, and contributing to various forms of analysis.

Green bottle flies play a crucial role in nature as essential scavengers, breaking down organic matter. They are commonly found near garbage and are the first insects to arrive at a carcass, laying eggs in decomposing matter. In forensic investigations, these flies are a valuable tool, with forensic entomologists often using maggots from green bottle flies to determine the time of death and treat non-healing wounds in humans.

Understanding the life cycle of green bottle flies is vital to appreciating their role in the ecosystem. After hatching, the larvae, spending up to three days eating decomposing animal matter, demonstrate their role as nature's recyclers. They reach maturity within two to 10 days and pupate in the soil, a testament to their Adaptability. They continue the cycle once they emerge as adults, mating and perpetuating their species. In cold weather, pupae and adults can hibernate until warmer temperatures return, showcasing their ability to survive in diverse conditions.

Understanding the diverse world of Diptera in your garden is essential for creating a balanced ecosystem. While many flies are often seen as pests, it's important to remember the crucial roles they play as pollinators, predators, and decomposers. By providing a variety of native plants, water sources, and undisturbed areas, you can encourage beneficial flies while minimizing the impact of nuisance species. Remember, a healthy garden is a balanced garden, and Diptera are an integral part of that balance.

Planting for Diptera

Discover Native Ontario Plants

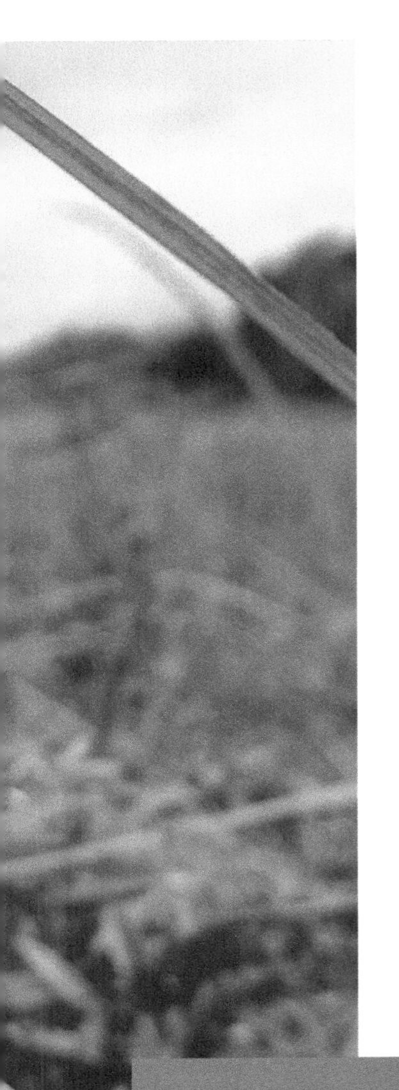

To create an irresistible haven for beetles in your garden, strategically planting native Ontario species goes beyond mere horticulture—it's an invitation to witness a thriving, interconnected ecosystem. Here's a closer look at some native Ontario plants from Ontario Native Plants that we suggest you, as gardeners, plant and their role in enticing and supporting the diverse world of beetles and other wildlife.

By strategically integrating these native Ontario plants and others into your garden, you're not just adding to its visual appeal but crafting a vigorous stage where beetles play pivotal roles. This carefully curated selection becomes an essential component of your garden's biodiversity, supporting beneficial beetle species and enhancing the overall ecological resilience of your outdoor space. As these plants bloom and flourish, they beckon beetles, turning your garden into a living tapestry where each species contributes to the intricate dance of life, from the ground-dwelling protectors to the aerial pollinators. Your actions are crucial in this delicate balance.

Wild Lupine
Lupinus perennis

◇◇◇◇◇◇◇◇◇◇◇◇◇◇◇◇

Growing Habits:

Mature Height: 2 feet
Mature Spread: 1 foot

Ontario Hardiness Zone: 3 to 7

Part Shade to Full Sun

Dry to Medium

Sand Loam

The Wild lupine (*Lupinus perennis*) is a captivating and visually striking plant that graces the landscape with its remarkable columns of elegant bluish-purple flowers. This perennial legume, with its distinctive palm-shaped leaves, showcases its beauty throughout the year. It thrives in sandy, well-drained soils, exhibiting resilience in environments with low nutrient and moisture levels.

Not only does the Wild lupine contribute to the visual aesthetics of its surroundings, but it also plays a crucial ecological role. It is the primary food source for the caterpillars of the rare and endangered Karner blue butterfly (*Lycaeides melissa samuelis*). By supporting the survival and reproduction of this butterfly, the wild lupine helps to maintain biodiversity and ecological balance in its habitats.

Furthermore, the Wild lupine exhibits an impressive ability to fix nitrogen, a valuable trait that contributes to soil fertility and health. This makes it a sought-after choice as a cover crop to enhance soil productivity and sustainability in agriculture.

Wild lupine, a treasure trove of nectar, is a haven for flies in the Diptera order. These flies, with their unique role as inadvertent pollinators, are crucial to the successful reproduction of the lupine plants. They are lured by the lupine's colourful, elongated flower clusters, which are rich in nectar. As the flies partake in the nectar, they unknowingly transfer pollen grains between flowers, promoting successful pollination for the lupine plants. This mutually beneficial relationship is a testament to the vital role of the Diptera species in the ecosystem.

Dense Blazing Star

Liatris spicata

◇◇◇◇◇◇◇◇◇◇◇◇◇◇

Growing Habits:

Mature Height: 5 feet
Mature Spread: 1 foot

Ontario Hardiness Zone: 3 to 7

Part Shade to Full Sun

Medium to Wet

Sand Loam, Loam

This tall, showy, pink-purple bloom, a true marvel of nature, lasts most of the summer. On their journey, they witness the enchanting sight of monarch butterflies and insects drawn by its sweet nectar. With its unique beauty, this plant requires more moisture than the shorter Cylindrical blazing star (*Liatris cylindracea*). However, it is an excellent choice for full-sun locations, adding a vibrant touch to your garden. *Liatris spicata* is not just a plant; it's a work of art that can grace your home with its stunning blooms for up to two weeks in a vase, a testament to its endurance and beauty—no wonder it's a popular choice among florists. *Liatris spicata* is a beautiful and versatile plant that can breathe life into any garden or landscape, enhancing its beauty and attracting beneficial insects.

Liatris spicata, with its profusion of diminutive flowers, acts as a vital support system for Diptera. These densely clustered blossoms serve as a bountiful reservoir of nectar, a vital sustenance for numerous fly species. As the flies traverse from one flower to another, drawn by the nectar, they inadvertently facilitate the transfer of pollen, thereby fostering pollination and ensuring the perpetual prosperity of the blazing star.

Canadian Elderberry
Sambucus canadensis

◇◇◇◇◇◇◇◇◇◇◇◇◇◇◇

Growing Habits:

Mature Height: 10 feet

Ontario Hardiness Zone: 3 to 7

Part Shade to Full Sun

Medium to Wet

Loam, Organic

The Canadian (American) elderberry (*Sambucus canadensis*) is a highly adaptable deciduous shrub indigenous to North America. It typically flourishes in moist or wet soil and requires ample sunlight, although it can tolerate partial shade. Its aesthetic appeal and reputation for producing edible berries have made it a sought-after plant in gardens and landscapes.

The elderberry's clusters of dark purple berries are well-known for their culinary applications, often used in making jams, jellies, pies, and even wine after being adequately cooked. However, it's essential to note that consuming the berries raw can be toxic due to cyanogenic glycosides.

Regarding landscaping, *Sambucus canadensis* is particularly suitable for expansive, low-lying areas within a garden or yard. Furthermore, its ability to propagate through suckers makes it ideal for naturalizing large areas. It is a versatile and beneficial plant for ornamental and practical purposes.

Sambucus canadensis offers a helping hand to many beetle varieties. The elderberry serves as a buffet for herbivorous beetles, providing tasty meals in the form of leaves, flowers, and even fruits. Detritivores, on the other hand, find a different kind of bounty. They feast on decomposing elderberry parts like fallen leaves and dead branches, aiding in the breakdown of organic matter and returning valuable nutrients to the soil. Beyond food, the elderberry provides a haven for beetles. The leaves and flowers offer shelter from predators and harsh weather, while some species even utilize the hollow stems of elderberry bushes as nurseries for their young. However, it's worth mentioning that this beneficial relationship is sometimes one-sided. While some beetles find the elderberry a helpful resource, others might see it as a tasty target, becoming pests themselves.

Black-Eyed Susans

Rudbeckia hirta

◇◇◇◇◇◇◇◇◇◇◇◇◇◇◇

Growing Habits:

Mature Height: 2 feet
Mature Spread: 3 feet

Ontario Hardiness Zone: 3 to 7

Part Shade to Full Sun

Dry to Medium

Sand Loam, Loam, Clay Loam

Black-Eyed Susans (*Rudbeckia hirta*), also known as Yellow daisies or Brown-eyed Susans, are hardy, perennial wildflowers native to North America. These beautiful flowers are well-adapted to various environments, including disturbed and unfavourable conditions, making them a resilient and sustainable choice for any garden or landscape. They typically bloom from mid-summer to early fall, adding a vibrant yellow colour to their surroundings.

One exciting feature of Black-Eyed Susans is their biennial blooming pattern, where they flower every two years. However, due to their prolific reseeding capabilities, they often appear to bloom annually. This natural reseeding behaviour makes them low-maintenance and reliable, ensuring a consistent display of their cheerful blooms in the garden.

Black-Eyed Susans make excellent companions to a wide range of other wildflowers, offering a beautiful contrast to species such as the Purple coneflower (*Echinacea purpurea*) and Wild bergamot (*Monarda fistulosa*). They also pair strikingly with ornamental grasses, such as Indian Grass (*Sorghastrum nutans*) and Little Bluestem (*Schizachyrium scoparium*), adding texture and visual interest to garden landscapes.

In addition to their aesthetic appeal, Black-Eyed Susans are valued for attracting pollinators, such as bees and butterflies, making them a popular choice among gardeners who prioritize ecological diversity. Their resilience, beauty, and capacity to support local wildlife make them a versatile and sustainable addition to any garden or natural landscape.

Hairy Beardtongue
Penstemon hirsutus

◇◇◇◇◇◇◇◇◇◇◇◇◇◇◇◇

Growing Habits:

Mature Height: 2 feet
Mature Spread: 1.5 feet

Ontario Hardiness Zone: 3 to 7

Part Shade to Full Sun

Dry to Medium

Sand Loam

The Hairy Beardtongue, scientifically known as *Penstemon hirsutus*, is a delightful, easy-to-care-for perennial that thrives in various garden settings. It forms an attractive clump of lush foliage, perfect for creating visual interest in garden borders and mixed plantings. This charming plant produces striking tubular flowers in delightful shades of pink, adorning the garden with their vibrant hues throughout spring. Its blossoms are a visual delight and serve as a magnet for important pollinators such as bees, butterflies, and hummingbirds, injecting a lively and dynamic energy into the garden.

Compared to its close relative, the Foxglove Beardtongue (*Penstemon digitalis*), the Hairy Beardtongue boasts a more compact stature. This makes it an excellent choice for smaller gardens or for those who prefer container gardening. Its manageable size and delightful blossoms make it a versatile and eye-catching addition to any landscape.

One of the critical attributes of the Hairy Beardtongue is its resilience and adaptability. Once established, this plant exhibits remarkable tolerance to cold and drought conditions, demonstrating its ability to endure and thrive in various environmental circumstances. This makes it an ideal candidate for gardeners who seek low-maintenance yet steadfast plant varieties for their outdoor spaces. Whether you're a seasoned gardener or just the Hairy Beardtongue, it will surely bring beauty, charm, and ease to your garden with its stunning display and reliable nature.

By incorporating these native plants into your Ontario garden, you're creating a vital habitat for a diverse range of Diptera. These often-overlooked insects play crucial roles in pollination, pest control, and nutrient cycling. Providing them with the right plants will not only enhance your garden's biodiversity but also contribute to the overall health of your local ecosystem. Remember, a garden teeming with life is a joy to behold, and these native plants are the foundation for a thriving Diptera community.

Hemiptera

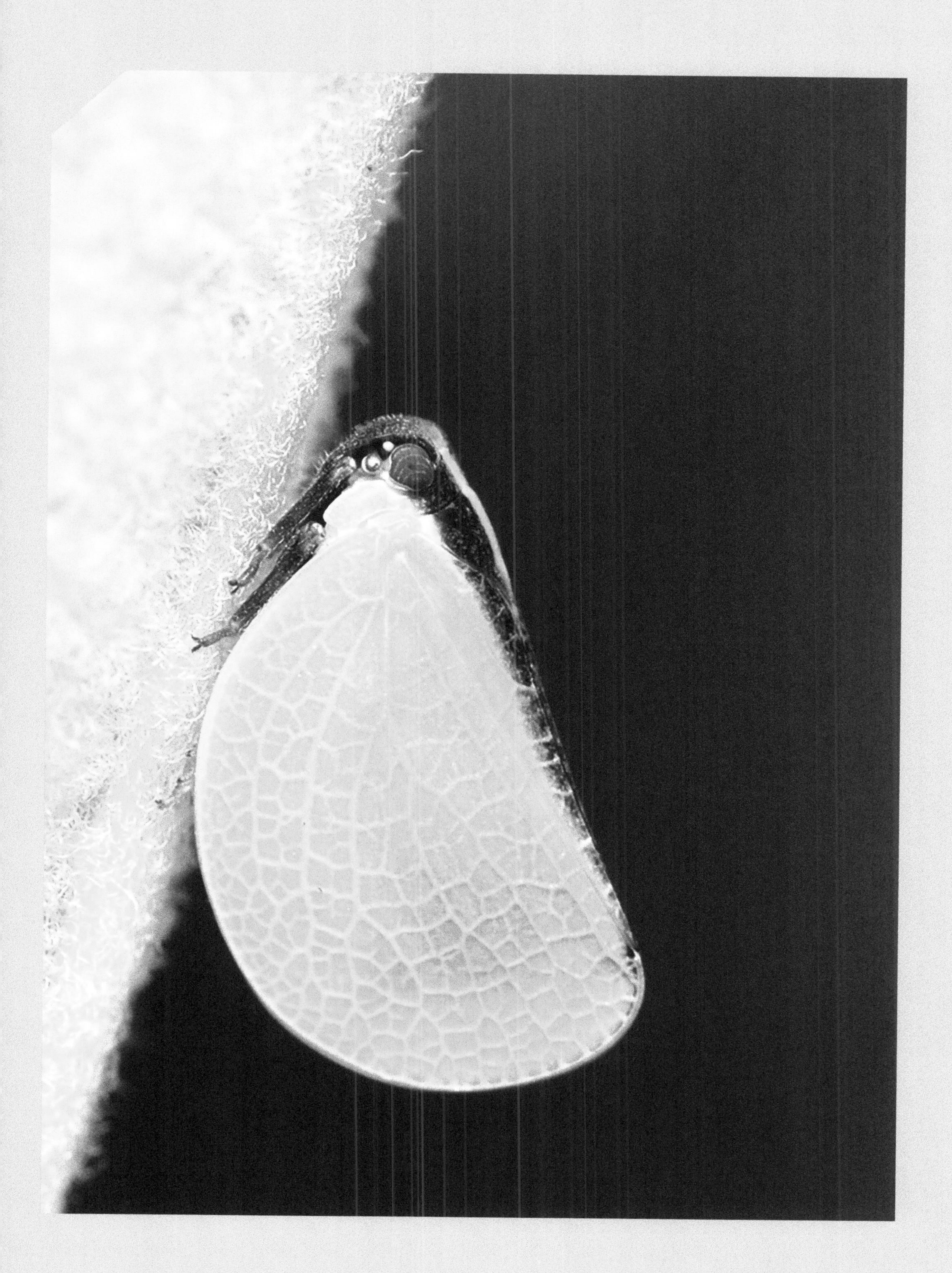

The Secrets of Ontario's True Bugs

◇◇◇◇◇◇◇◇◇◇◇◇◇◇

Step into your garden and behold a world beyond the vibrant blossoms. Beneath the petals, a hidden realm comes to life, orchestrated by nature's architects, the Hemiptera. Our gardens are not just canvases of colour but bustling ecosystems teeming with life and interwoven stories. At the heart of this tapestry are the unsung heroes, the Hemiptera, whose intricate lives contribute significantly to the ballet of our natural world.

These true bugs, a captivating Order of insects, wield specialized mouthparts, each species possessing a unique tale of survival and adaptation. With their specialized mouthparts, true bugs pierce and suck nectar from plants and body fluids from other insects, playing a significant role in the intricate web of life. As we delve into the heart of Ontario Gardens, we embark on a journey that transcends the ordinary, unravelling the secrets of over one thousand Hemiptera species, the architects, pollinators, and predators that balance nature's delicate dance.

This chapter will explore how every leaf and stem becomes a portal to a miniature universe, where the persistent embrace of aphids on a young shoot and the stealthy hunt of assassin bugs lurking in the shadows shape the vibrant canvas of your garden. We invite you to peer beyond the petals and discover a realm where every puncture and sip shapes the vibrant canvas of your garden.

In the following pages, we will traverse the diverse landscapes of Ontario's gardens, revealing Hemiptera's multifaceted roles in this living canvas. It is a journey of understanding, appreciation, and a call to coexist harmoniously with these enchanting insects. So, let the curtain rise and immerse ourselves in the captivating world of Hemiptera, where the ordinary becomes extraordinary. The microcosm beneath the leaves beckons us into a realm of fascination and awe.

Large Milkweed Bug,
Oncopeltus fasciatus

Top Right: *Lasiomerus annulatus*

Top Left: Coppery Leafhopper, *Jikradia olitoria*

Bottom: Say's Cicada, *Okanagana rimosa*

Unveiling Nature's True Bugs

◇◇◇◇◇◇◇◇◇◇◇◇◇◇

Picture waking up to a tranquil garden, the air filled with the sweet fragrance of flowers and a gentle breeze promising a beautiful day. As you take in the intricate patterns of plants and animals, consider the often-overlooked Hemiptera, an Order encompassing over 80,000 fascinating individual species, each playing a unique role in the garden's tapestry.

The story of Hemipterans begins with their unique and distinctive mouth parts, a marvel of evolution engineered for specific purposes. Unlike chewing insects like caterpillars, true bugs are characterized by their piercing-sucking mouthparts formed from modified needle-like stylets. These stylets can penetrate plant tissues and inject enzymes that liquefy the sap, allowing the Hemipteran to drink it. This remarkable adaptation fuels their diverse lifestyles, with some species, like cicadas, even possessing specialized filters to select specific nutrients from the sap. The sheer diversity of their lifestyles, from delicate sap feeders to specialized nutrient selectors, is truly amazing.

Hemiptera showcases an unparalleled variety, from tiny aphids delicately inserting their proboscides (modified mouthparts) into plant veins to massive water bugs exceeding 15 centimetres (6 inches) in length, thriving in aquatic environments. These water bugs, also known as giant water bugs, are not only impressive in size but also fierce predators, using their powerful legs to capture prey like tadpoles, small fish, and even other insects. This predatory behaviour is crucial in maintaining the balance of the ecosystem, a contribution that we can all appreciate.

What sets Hemiptera apart from other insects is their physical features and their crucial role in the intricate web of life within your garden. Visualize them as skilled musicians, each contributing their unique notes to a symphony that sustains the delicate balance of an ecosystem. Some Hemipterans, like aphids and plant hoppers, feed on plant sap, acting as herbivores. However, their presence is only sometimes detrimental. Ladybugs and assassin bugs, for example, are beneficial predators within the Hemiptera order. They actively hunt other insects, including those that can harm your precious plants, promoting a natural balance and acting as biological pest control.

As you delve into these pages, you'll uncover the captivating world of Hemiptera, exploring their anatomy, appreciating their ability to adapt to their environment, and understanding their interactions within gardens. From the fascinating social lives of aphids to the complex communication methods of cicadas, prepare to be amazed by the hidden wonders of these often-underestimated creatures.

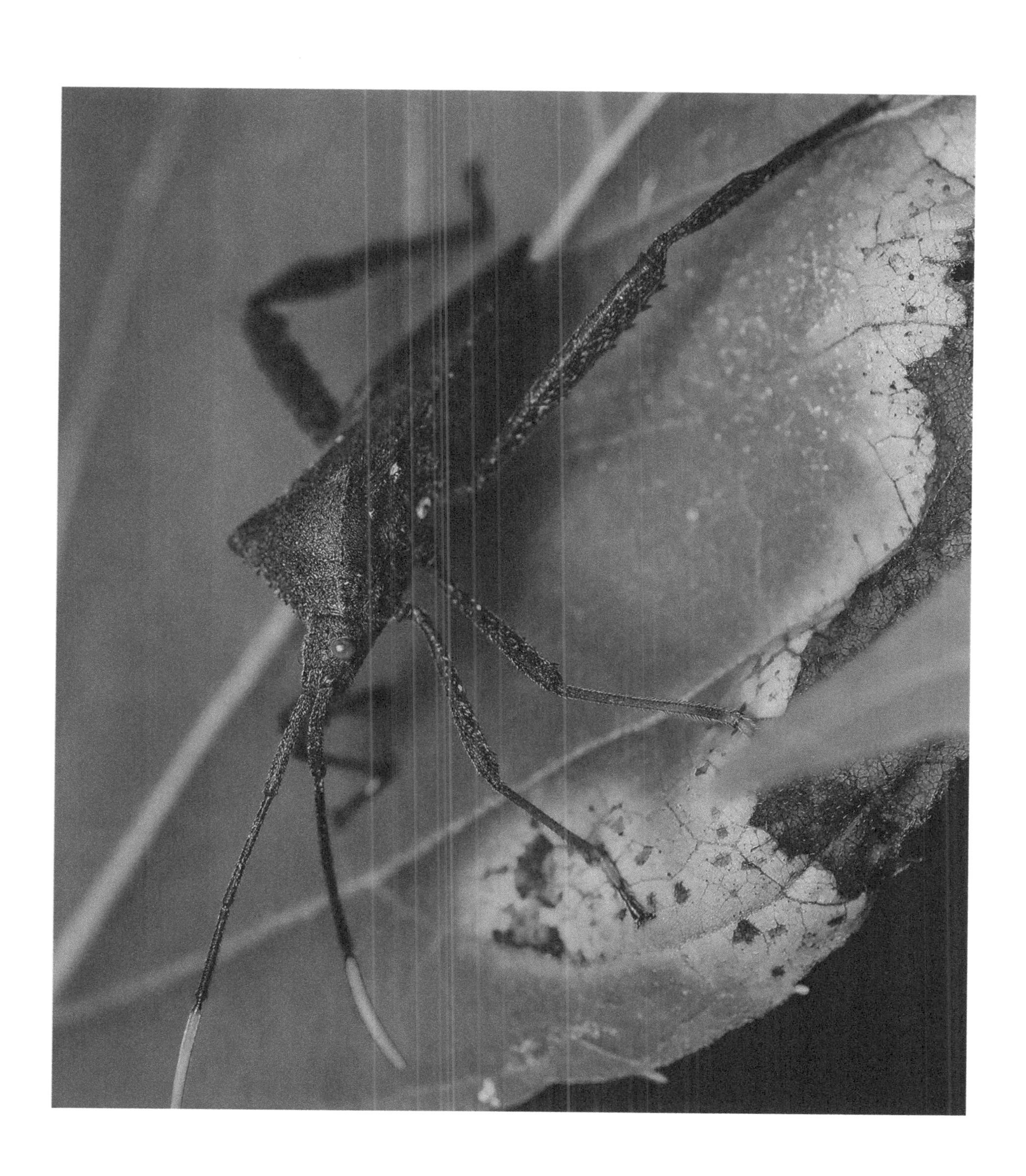

Acanthocephala terminalis

Why Do We Need Hemiptera?

◇◇◇◇◇◇◇◇◇◇◇◇◇◇

Hemiptera have adapted to exploit a wide range of different ecological niches by developing piercing-sucking mouthparts. In the garden landscape, these bugs have become skillful ecosystem engineers ranging from assassin bugs and predators to sap-sucking aphids. The predatory fractions of Hemipterans serve as nature's regulators, controlling common garden pests such as aphids, caterpillars and mites. This way, Hemipterans ensure that the natural predator-prey balance is maintained, thus reducing the need for chemicals while promoting a healthier natural environment suitable for plant growth.

Many Hemipteran insects act as predators and pollinate. For example, two well-known members of the order Hemiptera, cicadas and leafhoppers, travel through gardens searching for sap or honeydew while accidentally aiding pollen movement among flowers. Understanding that these insects play double roles as pest controllers and pollinators emphasizes the need to maintain a prosperous and thriving population of these insects in our gardens. Through this dual functionality, plant pests can be prevented while enhancing flowering species' reproductive success.

Their evolutionary success is evidenced by how adaptable Hemiptera are. These insects have developed various strategies to navigate and utilize different ecological niches. For instance, Shield bugs (*Fanmily Pentatomidae*) are renowned for their varied diet, feeding on both pests and plant saplings. By understanding subtle feeding habits and behaviour patterns among these bugs, gardeners can provide an informed approach to creating a garden with a balanced, resilient insect community.

Assassin bugs (*Family Reduviidae*) thrive across Ontario gardens, as they are the unheralded heroes who act as gardeners' dynamic and effective allies in pest control. The tiny killers that look like nature's slim assassins have long bodies accompanied by shiny forewings. They have long protrusive mouth parts for piercing soft-bodied insects like caterpillars, aphids and beetles. Imagine these vigilantes making their rounds over your flower beds, vegetable gardens, or herb gardens, taking out those bug nuisances whose presence threatens your botanical haven. In this way, assassin Bugs blend effortlessly into the garden ecosystem through cryptic coloration and agility hunting techniques, ensuring a balanced environment full of life.

Top: American Giant Water Bug, *Lethocerus americanus*

Middle: North American Tarnished Plant Bug, *Lygus lineolaris*

Bottom: Twice-stabbed Stink Bug, *Cosmopepla lintneriana*

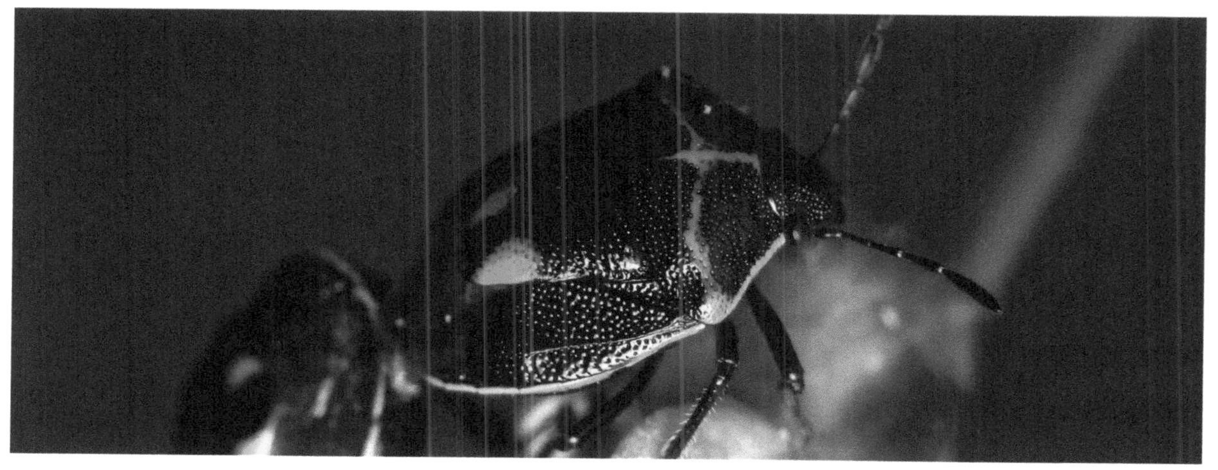

Morphological Marvels
Anatomy of Hemiptera

◇◇◇◇◇◇◇◇◇◇◇◇◇◇

One characteristic that distinguishes Hemiptera as true bugs involves their specially adapted mouthparts for piercing and sucking. The elongated proboscis, or a rostrum, is the multipurpose feeding tube that enables them to access plant fluids, sap, or even blood from other organisms in some species. This feeding habit in gardens with many host plants makes Hemiptera pests an integral part of maintaining a balanced ecosystem.

In Hemiptera, the wings have a unique arrangement, where the forewings are often thickened at the base and thin towards the tips. These structures, called "hemelytra," play dual roles: fly and protect. They protect the vulnerable hind wings when they rest. They facilitate swift movement during flight, thus enhancing maneuverability through garden surroundings due to their membranous tips.

Bright and varied colorations for aesthetic and functional purposes are common among hemipterans. Some species of Hemipterans, such as Firebugs (*Pyrrhocoris apterus*), have bold aposematic warning coloration, often red and black, which appear to deter passerine birds. They employ bright coloration to discourage predators by indicating unpalatability or toxicity. In contrast, others use cryptic colours, enabling them to blend effortlessly with their environment and making their survival more probable. This visual variation beautifies the garden space while revealing how Hemiptera has responded to different ecological niches.

Additionally, reproductive behaviour also gives a further understanding of the biology of Hemiptera. Many species, such as Greenhouse whiteflies (*Trialeurodes vaporariorum*), have complicated courtship rituals involving vibrations and pheromones to ensure successful mating between members of the opposite sex. Their ability to adapt and survive under varying ecological conditions is reflected by these insects' wide diversity of life cycles, ranging from simple metamorphosis in some groups to gradual metamorphosis in others.

Hemiptera's anatomy provides evidence of the evolutionary genius of these true bugs. Hemiptera contributes to garden ecosystems ' ecological diversity and balance through their mouth parts, distinctive wings, coloration, and complex reproductive behaviour. By understanding how these insects are built, gardeners can appreciate them as challenges and partners in keeping a robust and beautiful garden. We can promote the coexistence of these insects by maintaining an ecosystem that appreciates their ecological importance, thus making our gardens more sustainable.

Green Shield Bug,
Palomena prasina

Palomena prasina life cycle: Egg, Larva, Adult.

Positives and Negatives
Welcoming Hemiptera in our Gardens

◇◇◇◇◇◇◇◇◇◇◇◇◇◇◇

Hemiptera is an Order of insects that encompasses a wide variety of over one thousand Ontario species with different behavioural traits, some of which can significantly affect the garden ecosystem. Gardeners might learn the benefits and drawbacks of having Hemiptera in their gardens to maintain a balanced, healthy environment.

The presence of Hemiptera in gardens can be a boon for gardeners. Many Hemiptera species, such as Assassin bugs (*Family Reduviidae*) and minute pirate bugs, are natural predators that feed on garden pests like aphids, caterpillars, and mites. By encouraging these beneficial insects, gardeners can reduce their reliance on chemical pesticides, leading to a healthier and more sustainable garden.

Pollination is another positive feature of Hemiptera. Species such as Plant bugs (*Family Miridae*) and Stink bugs (*Family Pentatomide*) participate in pollination when feeding upon flowers. Although they may not pollinate as efficiently as bees, their role in the garden should still be recognized for what contributes to diversifying kinds of pollinators.

However, it's not all positive. Some Hemiptera, like Aphids (*Family Aphididae*) and Leafhoppers (*Family Cicadellidae*), are sap-feeding bugs that can harm plants by consuming plant sap, leading to distortion, yellowing, or even the spread of plant diseases. Gardeners must monitor these bugs and take necessary steps to prevent significant infestations.

It is important to note that some Hemiptera, like Stink bugs, produce defensive odours when disturbed, and the smell can be somewhat unpleasant. Though these odours help gardeners protect their gardens from predators, they should take care of them while dealing with them.

Hemiptera play significant roles in controlling pests and pollinating garden ecosystems. Gardeners can take advantage of these benefits by encouraging the use of beneficial insects while monitoring and checking potential threats from specific nectar-ingesting kinds. Understanding the subtlety of Hemiptera in the garden can enable a gardener to create a balanced and ecologically sound environment.

Alder Spittlebug,
Aphrophora alni

Left: Red-banded Leafhopper,
Graphocephala coccinea

Bottom: Brown Marmorated Stink Bug (Larva),
Halyomorpha halys

Discover Hemiptera

Ontario's Rich Hemiptera Diversity

Ontario boasts a rich diversity of Hemiptera, with over one thousand recorded species. These include aphids, leafhoppers, cicadas, and the enchanting water striders. Each species plays a unique role in the local ecosystem, influencing the health and vitality of your garden. Here are some of the common Hemiptera across Ontario gardens.

Here Is a Closer Look At The Top Five Hemiptera In Ontario Gardens

Masked Hunter
Reduvius personatus

◇◇◇◇◇◇◇◇◇◇◇◇◇◇◇

Size: Adult range in size between 17-23 millimetres.

Description Tips: The adult Masked hunters are elongated, black or dark brown, and somewhat glossy. They have relatively small, ovoid heads with wide-set eyes and a recognizable 'neck.' The thorax is darker and shinier than the rest of the body and has two noticeable protuberances or nubs on the dorsal surface. The wings are held horizontally over the abdomen, have a leathery appearance, and overlap the posterior two-thirds. The body and appendages are sparsely covered with upright hairs.

Immature masked bugs are typically lighter in colour than adults. Still, their coloration is often not observable as they usually cover themselves with dust, lint, bits of insect exoskeleton, and various other small items. These materials are glued to the nymph by a sticky, cuticular excretion, making them a living dust ball. Both the adults and nymphs have short, three-segmented, thickened mouthparts used to pierce their prey. The mouthparts are curved down and backward, and when not in use, they are tucked into a groove between the front pair of legs.

Reduvius personatus are beneficial predators who work tirelessly as nocturnal feeders and spend the daylight hours in protected, dry locations such as under heat registers, cabinets and cupboards, and inside wall voids and attics. Indoors, they feed on any arthropod they can overpower, including home pests such as silverfish, booklice, house centipedes, millipedes, carpet and hide beetle larvae and adults, bed bugs, and overwintering insects such as multicoloured Asian lady beetles, western conifer seed bugs, and cluster flies. They also feed on insects that accidentally fly in from outdoors during the warm months.

Masked hunters often indicate another underlying pest issue, making them a valuable indicator for homeowners and pest control professionals. By finding and eliminating the pests they're feeding on, we can effectively control the population of these beneficial insects.

Northern Dog-Day Cicada
Neotibicen canicularis

◇◇◇◇◇◇◇◇◇◇◇◇◇◇◇

Size: Adult range in size between 27-33 millimetres.

Description Tips: *Neotibicen canicularis*, also known as Dog-day cicadas, are fascinating insects with a robust physique, delicate transparent wings, and widely set eyes. Their unique coloration, which varies from vibrant green to earthy brown, black, and white, provides excellent camouflage among the foliage of trees.

The *Neotibicen canicularis*, commonly called the Northern dog-day cicada, is a fascinating insect resembling oversized flies. These creatures are predominantly concealed within their natural habitat and are often heard but seldom seen, as they spend most of their adult lives high up in the treetops. Their mating calls, which can reach almost deafening levels, are a familiar sound on warm summer days. Interestingly, these cicadas spend their entire nymph stages living underground, where they feed on the roots of plants. As they near the end of their development, the nymphs emerge from the soil, climb a nearby tree or structure, and undergo a final moult to transform into winged adult cicadas.

The duration of their development varies among species, with Annual cicadas taking as little as two years to reach maturity, while Periodical cicadas (*Genus Magicicada*) may take as long as 17 years. Immature Northern dog-day cicadas live underground for a few years, whereas adult northern dog-day cicadas only live for about 5-6 weeks above ground. This short lifespan above ground is dedicated to finding a mate, reproducing, and the females laying eggs. After this brief but intense period, the adults die, and the life cycle begins anew as the eggs hatch and the nymphs start their journey underground.

American Giant Water Bug
Lethocerus americanus

◇◇◇◇◇◇◇◇◇◇◇◇◇◇◇

Size: Adult range in size around 50.8-76.2 millimetres.

Description Tips: Native to North America, *Lethocerus americanus* occupies a significant eco-logical niche as an apex predator in aquatic environments. The American giant water bug is a visually striking species, distinguished by its deep brown hue and eye-catching banded raptorial legs that protrude from the front of its body. These unique features, along with its distinct lack of long antennae, set it apart from other creatures. These aquatic predators are formidable hunters, often mistaken for ticks or cockroaches. Despite this aerial capability, they are primarily marine, relying on their ability to swim and dive for survival and hunting. Respiration is achieved through short, retractable siphons at the tip of the abdomen, allowing them to extract oxygen from the air-water interface.

The American giant water bug holds significant ecological importance. They play a crucial role in regulating the populations of both invertebrates and vertebrates in ponds and lakes, preventing the overpopulation of other aquatic organisms. Their presence is a testament to the delicate bal-ance of pond ecosystems. By controlling populations, the Giant water bug helps prevent prey species from overpopulating and disrupting the fragile equilibrium of the ecosystem. By control-ling prey populations, they help to maintain water clarity and reduce algal blooms. A healthy predator-prey relationship contributes to a balanced and thriving aquatic ecosystem; through pre-dation, *Lethocerus americanus* accelerates nutrient cycling. Organic matter is transferred from one trophic level to another when it consumes prey.

Lethocerus americanus is a keystone species in its ecosystem. Its role as a top predator, contribu-tion to nutrient cycling, and potential as an indicator species highlight its importance in maintain-ing ecological balance. Understanding this insect's environmental role, particularly its potential as an indicator species, is crucial for effectively conserving and managing aquatic habitats. Its study can provide valuable insights into the health of marine ecosystems.

Jagged Ambush Bug
Phymata americana

◇◇◇◇◇◇◇◇◇◇◇◇◇◇◇

Size: Adult range in size between 8-11 millimetres.

Description Tips: The Jagged ambush bug is a captivating and powerful garden predator. These small creatures possess a unique combination of angular, greenish-yellow, white, and brown bodies with miniature wings that reveal their jagged abdomens. Their forelegs, robust and muscular like the raptors of a praying mantis, are used to seize and hold prey. With short beaks, they pierce their prey and secret saliva to dissolve the insect's internal organs. The final segment of their antennae, slightly clubbed, adds to their unique features, making them a fascinating subject of study.

Ambush bug coloration is highly variable within species, encompassing a spectrum of hues, including gold, yellow, green, tan, brown, and white, often punctuated by dark markings. Sexual dimorphism is evident in coloration, with males typically exhibiting darker or more intricate patterns than females. The precise mechanisms underlying this colour variation remain elusive. Potential explanations include colour-changing abilities akin to chameleons or crab spiders, habitat selection based on colour matching, or ontogenetic colour shifts correlated with developmental stages and environmental cues. Additionally, temperature conditions during embryonic development may influence adult pigmentation.

Jagged ambush bugs are masters of stealth in open meadows and gardens, often seen on the flowers of prairie plants like goldenrod and aster, where their body colouring serves as perfect camouflage. As their name suggests, they wait patiently for suitable insects, including flies, tiny moths, beetle larvae, and other soft-bodied true bugs attracted to the plants they sit on. Notably, they can catch prey much more significant than themselves, such as bumblebees and butterflies, showcasing their impressive predatory skills. More importantly, their ability to catch such prey demonstrates their crucial role in maintaining the ecological balance, making them an integral and admirable part of the ecosystem. Ontario is home to two common species of Jagged Ambush Bugs: Phymata Americana and Phymata Pennsylvania.

Pale Green Assassin Bug

Zelus luridus

◇◇◇◇◇◇◇◇◇◇◇◇◇◇◇

Size: Adult range in size between 12.5-18 millimetres.

Description Tips: The *Zelus luridus*, commonly known as the Pale green assassin bug, is a fascinating insect with a base colour of pale green. However, the markings on its back can range from dark brown or red to bright yellow, adding to its visual intrigue. The nymphs of this species are distinguishable by their solid green colour, lack of wings, and narrower bodies compared to the adults. One of the key identifying features of the *Zelus luridus* is the pair of spines on the rear corners of the pronotum, a plate-like structure covering part of the thorax. These spines are longer in lighter-coloured individuals and shorter in darker ones, aiding in the insect's camouflage and environmental adaptation. Furthermore, *Zelus luridus* can be identified by the presence of dark bands on the distal ends of the femurs, although these bands may sometimes be faint and challenging to discern.

Zelus luridus, commonly known as the Pale green assassin bug, is a fascinating insect species in deciduous trees and shrubs in forested and residential areas. These remarkable predators play a crucial role in the ecosystem by preying on various insects that make their home on the leaves of deciduous trees, shrubs, and different plants.

Assassin bugs belong to the *Reduviidae Family* and are skilled hunters. They exhibit patient hunting behaviour, often lying in wait on a leaf or stem to ambush their unsuspecting prey. However, they are also capable of actively seeking out their next meal.

The primary targets of the Pale green assassin bug are small flies, wasps, sawflies, and sedentary insects like caterpillars. Their hunting strategy involves patiently stalking and capturing their prey, contributing to regulating insect populations in their habitat.

The female Pale green assassin bug lays her eggs in clusters of around two dozen, often at the tips of leaves. These clusters are held together by a sticky, brownish substance, providing protection and support for the developing eggs. Once hatched, the nymphs of these fascinating insects are strikingly different from their adult counterparts. They are narrower-bodied, wingless, and entirely green in colour, eventually undergoing a remarkable transformation as they mature into adult assassin bugs.

By understanding the diverse world of Hemiptera and the important roles they play in your garden, you can create a thriving ecosystem. Remember, not all Hemiptera are pests; many are beneficial predators or pollinators. By providing a variety of native plants, you can attract a diverse range of these fascinating insects. Observe your garden closely, and you'll discover the intricate web of life supported by these often-overlooked creatures.

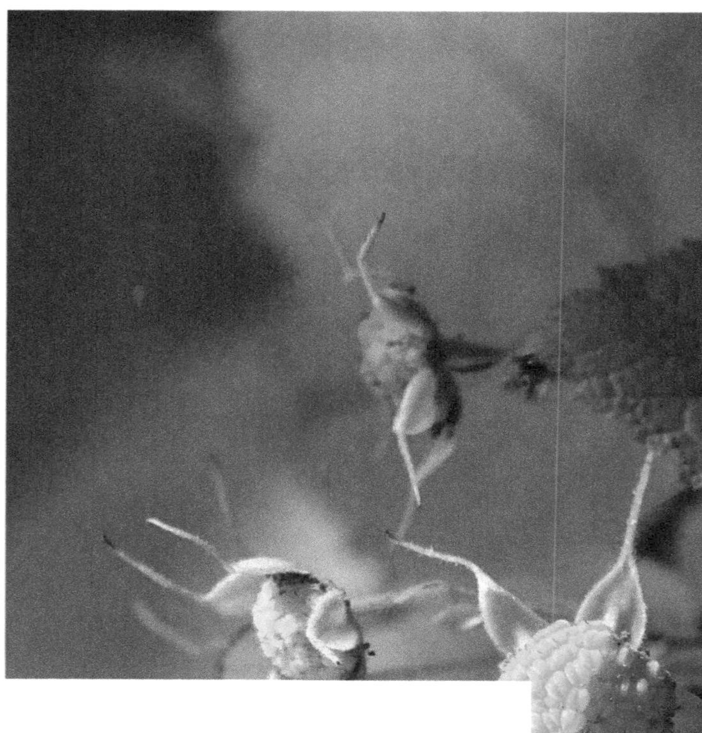

Planting for Hemiptera

Discover Native Ontario Plants

To transform your garden into a thriving haven for Hemiptera, consider planting a carefully curated selection of native Ontario plants. These plants not only enhance the beauty of your green space but also provide essential resources for these fascinating insects, encouraging their presence and contributing to the overall biodiversity of your garden. Here's a closer look at some native Ontario plants from Ontario Native Plants that we suggest gardeners plant and their role in enticing and supporting the diverse world of Hemiptera.

Your garden's Hemiptera interactions are an intricate web of ecological dynamics. As gardeners, you have the power to take proper action by recognizing that these insects can act as both pests and beneficial organisms for natural pest control, pollination, and nutrient cycling. Your appreciation of the wide range of Hemiptera species not only improves the gardening experience by offering aesthetic pleasure but also fosters a closer relationship with the fine-tuning of nature. In doing so, you adopt and manage the effect of Hemiptera, thus actively contributing towards creating a resilient and balanced ecosystem that sustains your plants and supports a myriad of life forms thriving in your gardens.

Thimbleberry
Rubus parviflorus

◇◇◇◇◇◇◇◇◇◇◇◇◇◇◇

Growing Habits:

Mature Height: 6 feet

Ontario Hardiness Zone: 3 to 7

Full Shade to Full Sun

Dry to Medium

Sand Loam, Loam

Rubus parviflorus, a name that belies its grandeur, surprises with its unique features. Large, fragrant white flowers, a sight to behold, bloom in captivating clusters from late spring through mid-summer. These blossoms transform into vibrant red berries in the summer, larger than raspberries (up to ¾ inch in diameter). Unlike raspberries that detach from the core, thimbleberries have a hollow core that remains attached when picked. The thornless stems, reminiscent of Purple flowering raspberries, make harvesting a breeze.

The delightful surprise continues with the flavour. Thimbleberries boast a tangy taste, sometimes with hints of citrus or musk, distinctly different from the sweetness of raspberries. They're perfect for enjoying fresh or turning into delectable jams and jel-lies, offering a unique culinary twist. But their value extends beyond the kitchen. This shrub provides vital seasonal sustenance for birds and mammals, becoming a welcome addition to your wildlife-friendly garden.

Large, maple-like leaves (up to 8 inches across) add a lush presence throughout the growing season, culminating in a stunning display of orange and maroon hues in the fall. Reaching about 5-6 feet tall and 3-5 feet wide, this low-maintenance shrub thrives in shaded areas with moist, well-drained soil, mimicking its natural habitat on woodland edges. This makes it an excellent addition to any garden's shrub border, offering beautiful blooms, delicious fruits, stunning fall foliage, and wildlife support – all in one package!

Dogtooth Daisy
Helenium autumnale

◇◇◇◇◇◇◇◇◇◇◇◇◇◇◇

Growing Habits:

Mature Height: 6 feet
Mature Spread: 3 feet

Ontario Hardiness Zone: 3 to 7

Full Sun

Medium to Wet

Sand Loam, Loam, Clay Loam

Scientifically known as *Helenium autumnale*, this plant is a remarkably abundant bloomer, gracing gardens with its charming yellow, button-shaped flowers that persist for weeks during late summer and fall. This extended blooming period is especially appreciated as it coincides with the waning of many other perennial blooms. While it has earned the rather unusual moniker "Sneezeweed" due to the historical use of its dried leaves in the production of snuff, which was used to induce sneezing for the purported purpose of expelling "evil spirits" from the body, it is more commonly referred to as Dogtooth Daisy.

Dogtooth Daisies are particularly well-suited to moist environments and exhibit a high tolerance for water, making them an ideal candidate for inclusion in rain gardens. It is frequently encountered in the wild near ponds or streams. As a taller wildflower, it is ideally suited for placement at the rear of a garden bed. Pairing it with Joe pye weed (*Eutrochium purpureum*), another tall species flourishing in moist conditions, is a favourable choice for creating a visually captivating and harmonious garden design.

Helenium autumnale is a beneficial plant for Hemiptera, like aphids and leafhoppers. These insects are attracted to the daisy's bright yellow flowers, where they feed on the nectar and pollen. While some may view them as pests, these Hemiptera play a role in the daisy's ecosystem by aiding in pollination.

Buttonbush
Cephalanthus occidentalis

Growing Habits:

Mature Height: 8 feet

Ontario Hardiness Zone: 4 to 7

Part Shade to Full Sun

Medium to Wet

Sand Loam, Loam, Clay Loam, Organic

Buttonbush, scientifically known as *Cephalanthus occidentalis*, is an exceptional choice for gardeners seeking to enhance low-sunny areas or pond surroundings with a stunning and functional shrub. This adaptable plant thrives in moist locations and can be planted directly in water, making it ideal for wetland gardens and naturalizing pond edges.

One of the most striking features of Buttonbush is its unique white, spherical flower heads, which bloom from A spring to early fall. These fragrant blooms attract pollinators, including bees, butterflies, and Hemiptera, making them an excellent addition to pollinator-friendly landscapes. In addition to its aesthetic appeal, buttonbush's ability to flourish in wet conditions also makes it a valuable option for stabilizing soil and preventing erosion around bodies of water. It's important to note, however, that buttonbush does not tolerate drought conditions well, so it's best suited for consistently moist or wet environments.

Cephalanthus occidentalis, a true bug magnet with its nectar-rich flowers and nutritious sap, is a key player in controlling pest insect populations. These features attract a variety of true bug species, including leafhoppers, aphids, and stink bugs. While some of these insects may feed on the plant itself, many others, like ladybugs, are beneficial predators that help maintain the ecological balance by controlling pest insect populations. Overall, buttonbush contributes to a healthy ecosystem by providing a food source for herbivorous and predatory Hemiptera.

Nannyberry
Viburnum lentago

◇◇◇◇◇◇◇◇◇◇◇◇◇◇◇

Growing Habits:

Mature Height: 15 feet

Ontario Hardiness Zone: 2 to 7

Part Shade to Full Sun

Medium to Wet

Loam, Clay Loam

A versatile native shrub with excellent year-round interest, Nannyberry features showy white flowers in May, followed by burgundy leaf color and dark blue berries in autumn. This large upright shrub can spread and form colonies, making it a good choice for a tall privacy screen or hedgerow. It can also be maintained as a small tree by pruning stems and removing the suckers at the base. Adaptable to a wide range of soils, this native viburnum is found in low moist woods or near stream banks, but will tolerate drier sites. Very shade tolerant in nature, it grows larger in open sunny areas.

The caterpillars of numerous small moth species are known to host on Viburnum lentago, which is one reason it is considered a top wildlife plant for nesting birds. Other interesting larval hosts include the Pink Prominent, the Hummingbird Clearwing, and the Green Marvel moth.

Viburnumns tend to flower profusely whether or not pollination occurs. However, poor fruiting will happen if there is only one Viburnum available. The edible berries can be used to make jams and jellies.

Wild Bergamot
Monarda fistulous

◇◇◇◇◇◇◇◇◇◇◇◇◇◇◇◇

Growing Habits:

Mature Height: 4 feet
Mature Spread: 3 feet

Ontario Hardiness Zone: 3 to 7

Part Shade to Full Sun

Medium

Sand Loam, Loam, Clay Loam, Organic

Monarda fistulosa, commonly known as Wild bergamot, is a versatile and delightful native wildflower that can be an excellent addition to any garden. This plant is a member of the mint family and features vibrant clusters of tubular, lavender-hued flowers, which are an absolute visual delight. Wild bergamot's aromatic foliage and flowers emit a sweet, citrusy fragrance that adds a delightful dimension to any garden setting. Not only are they pleasing to the eye and nose, but they are also edible, making them a versatile and valuable addition to any garden.

In addition to their aesthetic and aromatic appeal, Wild bergamot flowers are beautiful to a diverse array of pollinators, particularly butterflies, which adds a lively and colourful touch to your garden. Their nectar-rich blooms make them perfect for creating visually stunning bouquets and attracting beneficial insects to promote a thriving garden ecosystem.

Wild bergamot is adaptable, thriving in various soil types and moisture conditions, ranging from average to moist soils. It is known to flourish in anything from partial shade to full sun, making it an excellent choice for various garden settings. However, it's essential to be mindful when it comes to watering. Avoiding sprinkler or overhead watering systems can help prevent the development of powdery mildew on the plant, so it's best to water directly at the base of the plant to keep it healthy and looking its best.

By incorporating these native plants into your garden, you're providing essential habitat for a diverse array of Hemiptera. These insects, often overlooked, play crucial roles as predators, pollinators, and herbivores. Understanding their complex interactions within the garden ecosystem is key to creating a balanced and thriving space. Remember, many Hemiptera are beneficial insects that help control pest populations. By supporting their populations, you can reduce your reliance on chemical pesticides.

Hymenoptera

The Mysteries of Hymenoptera

◇◇◇◇◇◇◇◇◇◇◇◇◇◇

Step into the enchanting world of Ontario's gardens, where every delicate petal holds a mesmerizing tale and every verdant leaf whispers the secrets of a vibrant ecosystem. In this lush sanctuary, the Hymenoptera takes center stage, a diverse order of in-sects boasting over 1,300 species in Ontario alone. These masterful architects intricately weave beauty and essential threads into the intricate tapestry of nature's balance, connecting us to the larger web of life.

Picture a realm where the symphony of buzzing wings melds with the gentle rustle of leaves, and every creature, from the hum-ble bumble bee to the solitary mason wasp, plays a vital role in this ecological masterpiece. Observe the passionate dance of bees as they pollinate flowers, a process crucial for fruit and seed production in over 90% of flowering plants found in Ontario gardens. Witness the meticulous movements of ants, collaborating in complex societies with distinct castes (workers, soldiers, queens) forming the very foundation upon which the garden thrives. Some ant species cultivate fungus gardens within their colonies, further contributing to the ecosystem's decomposition process and nutrient cycling.

As we delve deeper into the captivating world of Hymenoptera, be prepared to be amazed by the marvels of these diminutive titans. Marvel at the architectural prowess of wasps constructing nests with intricate precision, from the papery nests of social wasps to the mud daubers who meticulously sculpt their abodes. Delve into the complicated networks of ant colonies, where teamwork and communication through pheromones form the bedrock of their success. These social insects can carry up to 50 times their body weight, a testament to their incredible strength.

Hymenoptera species transform Ontario's gardens into more than mere land-scapes; they shape them into flourishing theatres of biodiversity, inspiring awe and wonder. Embark on a journey where facts seamlessly intertwine with wonder, revealing the profound impact of Hymenoptera on the living canvas of our cherished gar-dens. The diligent work of these tiny creatures sustains the natural equilibrium, enriches the human experience, and offers a front-row seat to the awe-inspiring spectacle of nature's intricate design.

American Pelecinid Wasp,
Pelecinus polyturator

Top Left: Red-belted Bumble Bee, *Bombus rufocinctus*

Top Right: Blue Mud-dauber Wasp, *Chalybion californicum*

Bottom: Eastern Black Carpenter Ant, *Camponotus pennsylvanicus*

Why Hymenoptera Matter

◇◇◇◇◇◇◇◇◇◇◇◇◇◇

The Order Hymenoptera, which includes bees, wasps, and ants, is an essential and captivating part of the intricate ecosystems in Ontario's gardens. The name "Hymenoptera" comes from the ancient Greek words for "hymen," meaning membrane, and "pteron," meaning two wings, reflecting the characteristic two pairs of membranous wings of these insects. Globally, over a hundred and thirty thousand species of Hymenoptera have been identified, each playing a unique role in their ecosystems.

As crucial pollinators, bees have specialized adaptations such as pollen baskets and branched hairs for efficient pollen transportation. Their complex communication through dances facilitates collective foraging and resource optimization. With diverse lifestyles ranging from social colonies to solitary nesting, wasps contribute to pest control by preying on harmful insects. Ants, organized into complex societies, play multifaceted roles such as scavenging, seed dispersal, and soil aeration.

The life cycle of Hymenoptera is a fascinating process of complete metamorphosis, comprising distinct stages. It begins with the egg, laid in various locations depending on the species. The eggs hatch into larvae, voracious eaters with different forms adapted to their specific roles. A transformative process marks the pupal stage within a cocoon or nest, during which the insect undergoes significant anatomical changes. Finally, the adult Hymenoptera emerges, equipped with fully developed wings, specialized mouthparts, and distinct body structures. This metamorphic journey allows Hymenoptera to adapt to diverse ecological roles, contributing to the rich biodiversity of gardens and ecosystems.

Beyond their intriguing behaviours and characteristics, the significance of Hymenoptera within gardens extends to their paramount role as pollinators, ensuring the continued growth of diverse flora and sustaining the entire food web. Additionally, Hymenoptera regulates pests, helping maintain a natural balance by controlling populations of harmful insects. Their activities contribute to the health and resilience of garden ecosystems, fostering biodiversity and enhancing the overall well-being of the environment.

In essence, the presence of Hymenoptera in gardens is not just a matter of fascination but a crucial element in maintaining nature's delicate equilibrium. By understanding and appreciating these insects' roles, gardeners can actively contribute to fostering environments that support Hymenoptera, thereby promoting the health and sustainability of their gardens.

Great Golden Digger Wasp,
Sphex ichneumoneus

Why Do We Need Hymenoptera?

◇◇◇◇◇◇◇◇◇◇◇◇◇◇◇

The native bees, as pollinators, have been the backbone of Hymenoptera's significance in gardens. Bees have various structures for collecting and transferring pollen, which play a vital role in the reproduction process of many flowering plants. The vast number of native bee species ensures pollination among different kinds of plants, thus ensuring genetic variations and the entire stability of a garden system. Gardeners who create an environment that promotes the presence of native bees ultimately lead to vast numbers of fruits, vegetables and flowers hence healthful and productive gardens.

Besides their role as pollinators, some Hymenopterans are also efficient hunters and controllers of pests within gardens. An example is parasitoid wasps, which lay their eggs on or inside some pest insects like the caterpillars of the Five-spotted hawk moth (*Manduca quinquemaculata*), leading to the eventual death of the host. This natural form of biological control assists gardeners in managing pests without using chemicals. Understanding and appreciating predatory traits of specific hymenopteran species helps gardeners rely on them for a balanced and sustainable pest management plan. Ants, another critical group within Hymenoptera, help improve garden soil fertility and nutrient cycling. Their nest-making activity eases water infiltration into the soil while allowing absorption by plant roots. Ants also disperse seeds, thus aiding in plant regeneration. Viewing ants as beneficial contributors to our garden ecosystems lets us incorporate them into our ecosystems to enhance soil fertility and make plants more resilient through their ecological services.

The presence of Hymenoptera within garden landscapes contributes to a green space that is robust and sustainable. Bees, wasps, and ants have different roles in pollination, pest control and ecosystem engineering. Thus, they present an all-encompassing approach to gardening that goes beyond aesthetic concern about flowers and plants. Gardeners who support beneficial populations of Hymenoptera contribute to the well-being of their gardens, hence maintaining ecological balance and sustainability. Let us recognize that Hymenoptera is important in this natural environment and, through our stewardship, create a space that hums with the peaceful drone of pollinators, the vigilant hunt of wasp predators, and the bustling activities of ant colonies.

Pure Green Sweat Bee.
Augochlora pura

Hedgehog Gall Wasp,
Acraspis erinacei

Morphological Marvels
Anatomy of Hymenoptera

◇◇◇◇◇◇◇◇◇◇◇◇◇◇

The native bees, as pollinators, have been the backbone of Hymenoptera's significance in gardens. Bees have various structures for collecting and transferring pollen, which play a vital role in the reproduction process of many flowering plants. The vast number of native bee species ensures pollination among different kinds of plants, thus ensuring genetic variations and the entire stability of a garden system. Gardeners who create an environment that promotes the presence of native bees ultimately lead to vast numbers of fruits, vegetables and flowers hence healthful and productive gardens.

Besides their role as pollinators, some Hymenopterans are also efficient hunters and controllers of pests within gardens. An example is parasitoid wasps, which lay their eggs on or inside some pest insects like the caterpillars of the Five-spotted hawk moth (*Manduca quinquemaculata*), leading to the eventual death of the host. This natural form of biological control assists gardeners in managing pests without using chemicals. Understanding and appreciating predatory traits of specific hymenopteran species helps gardeners rely on them for a balanced and sustainable pest management plan. Ants, another critical group within Hymenoptera, help improve garden soil fertility and nutrient cycling. Their nest-making activity eases water infiltration into the soil while allowing absorption by plant roots. Ants also disperse seeds, thus aiding in plant regeneration. Viewing ants as beneficial contributors to our garden ecosystems lets us incorporate them into our ecosystems to enhance soil fertility and make plants more resilient through their ecological services.

The presence of Hymenoptera within garden landscapes contributes to a green space that is robust and sustainable. Bees, wasps, and ants have different roles in pollination, pest control and ecosystem engineering. Thus, they present an all-encompassing approach to gardening that goes beyond aesthetic concern about flowers and plants. Gardeners who support beneficial populations of Hymenoptera contribute to the well-being of their gardens, hence maintaining ecological balance and sustainability. Let us recognize that Hymenoptera is important in this natural environment and, through our stewardship, create a space that hums with the peaceful drone of pollinators, the vigilant hunt of wasp predators, and the bustling activities of ant colonies.

European Honeybee,
Apis mellifera

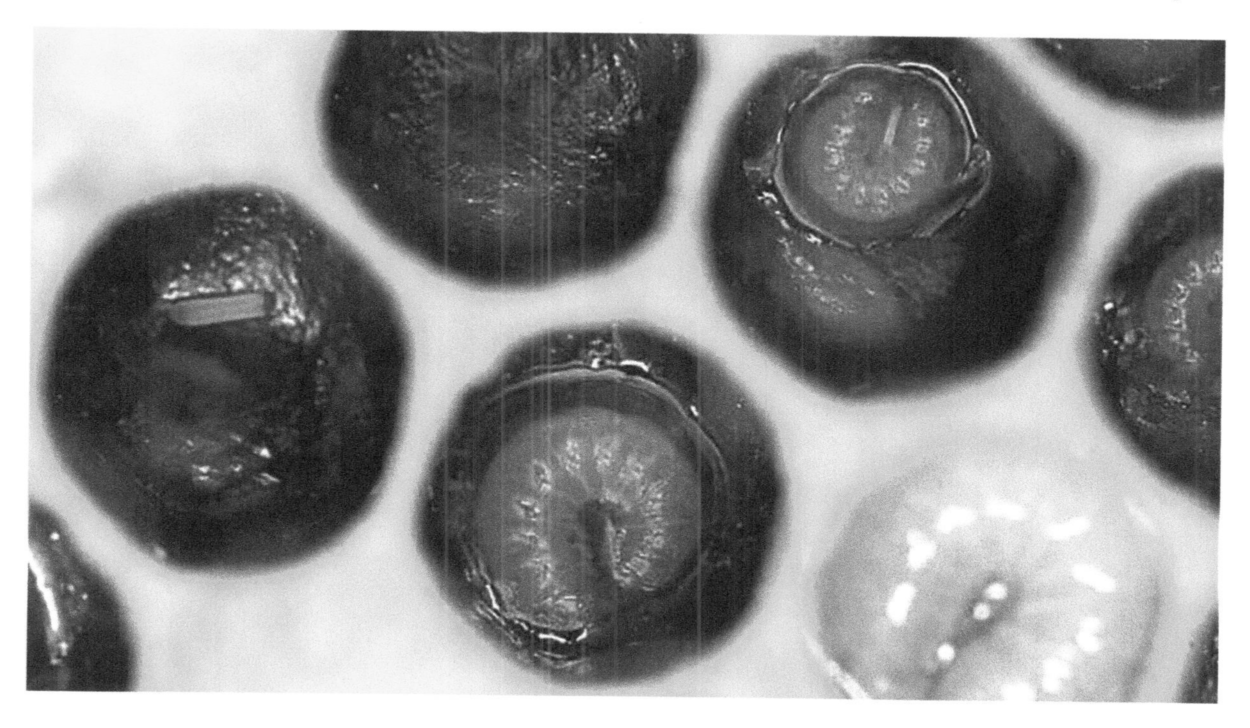

Apis mellifera life cycle, eggs, larva, pupa and adult.

Positives and Negatives
Welcoming Hymenoptera in our Gardens

◇◇◇◇◇◇◇◇◇◇◇◇◇◇◇

The Order Hymenoptera consists of bees, wasps, and ants, each critical in garden ecosystems. Although having Hymenoptera in your garden has advantages and disadvantages, these insects are pivotal in sustaining a happy balance between plants and insects.

One of the most essential advantages of Hymenoptera in the garden is its crucial role in pollination. Native bees are widely known and effective pollinators that help plants reproduce many species of flowering plants. Specialized relationships between native bee species and specific plants also increase the garden's biodiversity. Promoting a variety of bees contributes to robust pollination services, enhancing fruit and seed production.

Moreover, some wasp species belonging to Hymenoptera, like Parasitoid wasps, are natural predators of garden pests. Parasitoid wasps such as *Cotesia congregatus*, who deposit their eggs on Tomato hornworms (*Manduca quinquemaculata*) and others who are often caterpillars, such practices effectively control pest populations. Each parasitoid wasp species needs another species of insect to complete its lifecycle. Adults seek out other insects, such as caterpillars; they lay their eggs inside or on it when they find it. Thus, they can be helpful counterparts for gardeners looking for ways to prevent natural and organic pests.

However, some species of Hymenoptera have less attractive attributes. Social wasps like the German yellowjacket (*Vespula germanica*) and the Bald-faced hornet (*Dolichovespula maculate*) may be considered annoying pests, especially during late summer when the colonies are large enough that feeding drives them to seek food aggressively. However, potential issues with these wasps can be prevented through proper waste management and ensuring food is not attractive to them.

Ants like the common Citronella ant (*Lasius claviger*) play a positive role in aerating soils and recycling nutrients; however, some species may act as protectors of their mutualistic allies with pests like aphids that suck plant sap. Gardeners must observe ant-aphid interactions to save plants from possible damage.

Having Hymenoptera within your garden will have many advantages, such as adequate pollination and natural pest control. As a result, fostering diverse communities of native bees and using parasitoid wasps to control pests can contribute positively to the overall well-being of the garden. Gardeners who are aware of such potential threats as food-seeking social wasps or mutualistic connections between some ants and aphids can be prepared to approach the issue thoughtfully. Thus, they can contribute towards a balanced and environmentally sound garden.

Cherry Slug Sawfly (Larva),
Caliroa cerasi

Middle: Eastern Cicada-killer Wasp, *Sphecius speciosus*

Bottom: European Fire Ant, *Myrmica rubra*

Discover Hymenoptera

Ontario's Rich Hymenoptera Diversity

Ontario boasts a diverse array of over one thousand three hundred Hymenoptera species, showcasing a rich assortment of ecological importance. The industrious European honeybee (*Apis mellifera*) stands out for its social structure and pivotal role in pollination. Solitary wasps, such as members of the Sphecidae Family, exhibit extraordinary nesting behaviours, meticulously constructing individual nests to house their developing offspring. This diversity extends to the intricate world of ant colonies, with Formicidae species contributing to soil health, seed dispersal, and cooperative hunting strategies. These Hymenoptera, each with unique characteristics and ecological roles, collectively shape the complex and interconnected web of life within your backyard in Ontario. Here are some of the common Hymenoptera across Ontario gardens.

Here Is a Closer Look At The Top Five Hymenoptera In Ontario Gardens

Common Eastern Bumble Bee

Bombus impatiens

◇◇◇◇◇◇◇◇◇◇◇◇◇◇◇

Size: Adult range in size between 16-20 millimetres.

Description Tips: Bumble bees are vital for pollination because of their distinctive physical features. These fantastic creatures are densely covered in soft, fuzzy hair, perfect for collecting and trapping pollen as they move from one flower to another. The Common eastern bumble bee, a particularly unique species, has black hair on its head, abdomen, and legs. At the same time, its thorax is a striking shade of vibrant yellow, giving it a truly distinctive appearance.

Bombus impatiens, is a crucial pollinator due to its remarkable ability to transfer pollen from the male part of the flower (stamen) to the female part (stigma). These social bees live in colonies, with newly mated queens hibernating over the winter and emerging in early spring to find new nesting sites. The nests typically house 300 to 500 individual workers and are used annually. Towards the end of summer, males and new queens are born, and in the fall, worker bees, males, and the old queen die off. *Bombus impatiens* is the most commonly encountered bumblebee species in much of Eastern North America, and its range has recently expanded into Eastern Canada and parts of Western Canada. This species is frequently used for commercial pollination in greenhouses and outdoor crops.

Unlike many other insects, bumblebees can regulate their body temperatures by rapidly contracting their flight muscles. This behaviour allows them to forage for food during early spring temperatures and cold spells throughout their flight seasons. This unique ability is essential for their survival and productivity as pollinators.

European Honey Bee
Apis mellifera

◇◇◇◇◇◇◇◇◇◇◇◇◇◇◇

Size: Adult range in size between 11-18 millimetres.

Description Tips: *Apis mellifera*, commonly known as the honeybee, is a fascinating insect species that plays a vital role in our ecosystem. These bees are easily identifiable by their two sets of wings and black or brown bodies adorned with vibrant yellow bands on their abdomen. The colony consists of three main types of bees: the queen bees, the fertile females who are the largest; the medium-length, stout drones, the males, which have significantly larger eyes than the females; and the small sterile female worker bees, which have barbed stingers and unique hind legs that serve as pollen baskets, showcasing a fascinating adaptation. These distinctive physical characteristics contribute to the honeybees' effectiveness as pollinators, emphasizing the significance of their conservation efforts.

The European honey bee, is a tiny yet significant insect for the global food production ecosystem. These bees are critical in pollinating diverse valuable crops, including strawberries, grapes, apples, and almonds. It's important to note that many of these crops heavily depend on Honey bee pollination for successful fruit sets and high-quality yields. *Apis mellifera* is responsible for pollinating approximately one-third of humans' daily food. Similarly, Lepidoptera, such as butterflies and moths, play a crucial role in pollination and biodiversity. By understanding and appreciating these insects, we can better care for our gardens and the environment.

Honey bees are highly social insects that live and work together in large colonies known as hives, which can house up to 100,000 individual bees. Within these hives, a queen, the colony's leader, is responsible for laying eggs, while most bees are diligent workers. The queen Honey bee, with her astonishing ability to lay up to 1,500 eggs daily, is a testament to the growth and sustainability of the hive.

Honey bees diligently collect nectar and pollen from flowers and store them within the hive. The collected nectar, a sweet and valuable substance, is transformed into honey, a fascinating and essential process for bee survival. The pollen, conversely, undergoes fermentation to become bee bread, a crucial protein source for the bees. This intricate process of honey production and pollen fermentation highlights the bees' adaptability and resourcefulness in their environment.

Northern Paper Wasp
Polistes fuscatus

◇◇◇◇◇◇◇◇◇◇◇◇◇◇◇◇

Size: Adult range in size between 19-22 millimetres.

Description Tips: The Northern Paper wasp, known scientifically as *Polistes fuscatus*, is a marvel of nature. Its elegant and slender body, with a distinct waist that connects the thorax to the abdomen, is a testament to the diversity and beauty of the insect world. These wasps typically display a dark brown coloration, often adorned with lighter brown, coppery brown, or yellow markings, lending them a unique and striking appearance. One particularly intriguing feature of this species is the noticeable difference in forewing length between males and females. Males typically have forewings measuring above 13 millimetres, while females' forewings are generally around 11 millimetres in size, highlighting a subtle but fascinating distinction between the two genders.

Polistes fuscatus is a species commonly found in the eastern half of North America, extending from southern Canada to the United States and reaching as far north as Chilcotin, British Columbia. These fascinating wasps thrive in wooded forested areas and human-inhabited regions where they can find the resources necessary for constructing their nests. The construction of their nests involves using wood from their surroundings, which is chewed into a pulp-like material with the help of fluid from the wasps' mouths. Typically, these nests consist of a single paper comb without a protective wrapping and are often situated in sheltered locations, such as beneath overhanging structures.

One intriguing aspect of the behaviour of *Polistes fuscatus* queens is their tendency to invade and take over the nests of other queens. Additionally, only female paper wasps possess stingers, as these are modified ovipositors (egg-laying organs). On the other hand, male paper wasps are the last to be raised in a nest and primarily serve the purpose of mating with new queens. In terms of foraging, these wasps seek out nectar and soft-bodied insects like caterpillars, and their predation on such insects can significantly impact the local ecosystem by reducing the population of these essential creatures.

Eastern Black Carpenter Ant
Camponotus pennsylvanicus

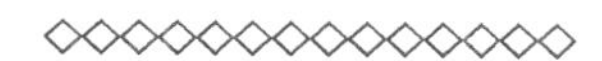

Size: Adult range in size between 5-18 millimetres.

Description Tips: The queen of *Camponotus pennsylvanicus*, also known as the eastern black carpenter ant, is distinguishable by her larger size than the worker ants. She has a robust thorax and wings, whereas most worker ants are wingless. Measuring about half an inch (1.2 centimetres) in length, these carpenter ants display a striking black to reddish-brown coloration and are adorned with yellowish hairs on their abdomens. Additionally, they are characterized by their prominent jaws.

Carpenter ants are a common sight in Ontario, Canada, especially the Eastern black carpenter ant, the most prevalent species in the region. These ants are known for their impressive size variation within species and colonies, making them some of the most giant ants encountered by gardeners. They are adept at nesting in wood, using their powerful jaws to create intricate tunnels within the wood. While they play a crucial role as natural decomposers in the wild by inhabiting dead and decaying logs and trees, they can also threaten structures when infested homes and businesses, particularly in moist or water-damaged areas.

Each *Camponotus pennsylvanicus* colony has a queen and numerous sterile workers responsible for brood care, nest defence, and food collection. These ants are nocturnal foragers, often spotted outside their hives while searching for small insects at night. When found indoors, carpenter ants typically search for food, as they do not consume wood but rather use it to construct their nests. Their diet primarily consists of plants, smaller insects, and a sugary substance aphids produce, which they cultivate and protect for their consumption.

Bicolored Striped Sweat Bee
Agapostemon virescens

◇◇◇◇◇◇◇◇◇◇◇◇◇◇◇

Size: Adult range in size between 6-8 millimetres.

Description Tips: *Agapostemon virescens*, a common native bee in Ontario, stands out with its distinctive metallic green head, thorax, and striped abdomen. Female bees sport white and black stripes, while their male counterparts display yellow and black stripes. However, some females may have a solid green abdomen. These bees, known as 'sweat bees,' are less likely than bumblebees to land on you. Instead, they can be found foraging on various flowers. Their short tongues make them fre-quent visitors to flowers with very short tubes that allow them to reach the nectar.

While the Bicolored striped sweat bee is primarily solitary, it sometimes exhibits communal nest-ing behaviour. In these instances, several females may share the same entrance tunnel but then build individual branches of the shared tunnel, with each branch having additional branches for the female's eggs. When the egg hatches, the larva will eat the ball of pollen that its mother left for it.

Agapostemon virescens plays a pivotal role in ecological systems as a primary pollinator. Its sig-nificant contribution to the reproductive success of various flowering plants is undeniable. By transferring pollen between flowers, A. virescens facilitates plant fertilization, leading to seed production and subsequent plant population maintenance and expansion. Moreover, their forag-ing behaviour can influence plant community composition and genetic diversity. Given their preference for open, sunny habitats and diverse floral resources, these bees indicate ecosystem health and can serve as valuable bioindicators. Understanding the ecological role of *Agapostemon virescens* is crucial for effective conservation and management strategies aimed at preserving pollinator populations and promoting plant biodiversity.

The world of Hymenoptera is incredibly diverse and fascinating. These insects play crucial roles in pollination, pest control, and nutrient cycling. By understanding their complex life histories and habitat requirements, you can create a garden that supports a thriving population of bees, wasps, and ants. Remember, many of these insects are beneficial and essential for a healthy ecosystem.

Planting for Hymenoptera

Discover Native Ontario Plants

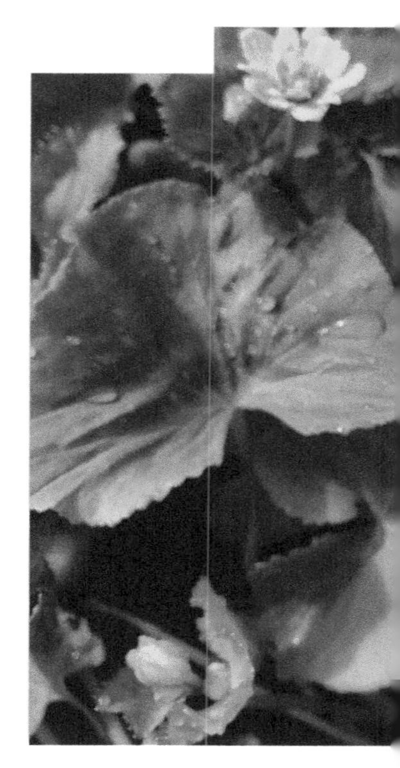

Your strategic planning is vital to attracting Hymenoptera to your garden. By planting native Ontario plants, you play a crucial role in supporting these critical insects. Here are some recommended native Ontario plants that you, as a gardener, can cultivate to draw and sustain the various types of Hymenoptera from Ontario Native Plants.

Establishing a sanctuary for Hymenoptera in your garden nurtures a thriving natural habitat and acts as a catalyst for the broader ecosystem's well-being. The presence of these pollinators is a testament to the intricate ecological symphony, enriching biodiversity and ensuring the robustness of various plant species. Witnessing the tireless efforts of Hymenoptera in pollinating flowers underscores the importance of preserving and promoting these vital components of the natural world, inspiring us to do our part.

Bottle Gentian
Gentiana andrewsii

◇◇◇◇◇◇◇◇◇◇◇◇◇◇◇

Growing Habits:

Mature Height: 2 feet
Mature Spread: 1 foot

Ontario Hardiness Zone: 3 to 7

Part Shade to Full Sun

Medium to Wet

Sand Loam, Loam

Gentiana andrewsii is characterized by its striking appearance, boasting uniquely shaped dark blue flowers that take on a bottle-like form and remain closed throughout the fall season. These flowers are particularly fascinating as bumblebees primarily pollinate them; the flower's structure allows only a few select insects to access its interior. This slow-growing plant is well-suited for those seeking low-maintenance greenery, and it flourishes in cool, acidic, and well-draining soils, especially in partially shaded areas. It is an excellent addition to the border of a woodland or shade garden. The plant's preference for moist soil also makes it an ideal choice for planting along streams or ponds, adding beauty and interest to these areas.

While preferring part shade, this easy-to-grow perennial thrives in moist, well-drained soils with rich organic matter and a slightly acidic pH. With its captivating blooms and manageable care requirements, the closed bottle gentian is a welcome addition to any wildflower haven.

Bottle gentian is a pollinator magnet, particularly for Hymenoptera. Its closed bell-shaped flowers hold a sweet nectar reward, but only the most vital of pollinators, the bumblebees, with their unique abilities, can reach it. The bumblebee forces its way inside the flower, getting dusted with pollen in the process, which it then transfers to other bottle gentian plants, promoting pollination.

Eastern Redbud
Cercis canadensis

◇◇◇◇◇◇◇◇◇◇◇◇◇◇◇

Growing Habits:

Mature Height: 30 feet

Ontario Hardiness Zone: 5 to 7

Part Shade to Full Sun

Medium to Wet

Loam, Clay Loam

The Eastern Redbud, also known as *Cercis canadensis*, is a captivating tree originally native to the southernmost part of Ontario in Canada. It is a popular choice for landscaping gardens and can be found as far north as Toronto despite not being native to natural areas in that region. This deciduous tree typically has a short trunk, a rounded crown, and gracefully spreading branches, creating a picturesque silhouette in the landscape.

One of the Eastern Redbud's most striking features is its distinctive heart-shaped leaves, which turn a beautiful yellow in the fall. This tree bursts into a stunning display of vibrant pink blossoms in the spring, adding colour to any garden. The flowers appear before the leaves, enhancing their visual impact. These delightful blossoms attract pollinators and bring a touch of elegance to the landscape.

In addition to its ornamental value, the Eastern Redbud also has practical uses. Its flowers and seeds are edible and can be enjoyed in various culinary creations. This adds a layer of value to an already charming and attractive tree. With its beauty, adaptability, and practical benefits, the Eastern Redbud is a versatile and valuable addition to any garden or landscape.

Spotted Joe-Pye Weed
Eupatorium maculatum

◇◇◇◇◇◇◇◇◇◇◇◇◇◇◇

Growing Habits:

Mature Height: 7 feet
Mature Spread: 3 feet

Ontario Hardiness Zone: 3 to 7

Part Shade to Full Sun

Medium to Wet

Clay Loam, Organic

Spotted Joe-Pye is a beautiful choice for enhancing your garden's beauty and ecological value. Its tall and robust stature, adorned with large clusters of pink flowers, serves as an alluring magnet for a diverse array of pollinators. This plant is adaptable and can thrive in various garden environments as long as the soil is consistently moist. By cultivating Spotted Joe-pye, you actively contribute to the conservation of native bees and butterflies, as it provides essential resources for these vital pollinators. Furthermore, the plant's hollow stems serve as valuable nesting sites for native bees, significantly promoting biodiversity in your garden.

Spotted Joe Pye weed is a magnet for Hymenoptera, the order that includes bees, wasps, and ants. These pollinators play a crucial role in the plant's reproduction, flocking to the plant's attractive clusters of purple flowers, feasting on the readily available nectar and pollen. Its attractiveness lies in its abundant nectar production throughout the summer months, a crucial energy source for these pollinators. The plant's flower structure, with easily accessible tubular blooms, caters to a wide range of insect sizes. As a native species, Spotted Joe-Pye weed supports local ecosystems and the insect populations that have evolved alongside it.

Marsh Marigold
Caltha palustris

◇◇◇◇◇◇◇◇◇◇◇◇◇◇◇

Growing Habits:

Mature Height: 1 foot
Mature Spread: 1 foot

Ontario Hardiness Zone: 3 to 7

Part Shade to Full Sun

Medium to Wet

Sand Loam, Loam, Clay Loam, Organic

An early-flowering beauty in the Buttercup family that is perfect for wet areas. Dark, waxy leaves contrast with vibrant yellow flowers appearing in early spring and persisting into early summer. Plants are clump-forming, creating attractive mounded groups. This plant occurs naturally along stream banks and pond edges where the soil is consistently moist to wet with part to full sun. *Caltha palustris* can survive occasional droughts but will lose their leaves and remain dormant. Some historical records claim that parts of the Marsh Marigold can be eaten. However, these plants contain toxic compounds and must be boiled in multiple water changes before consumption. Leaves are generally resistant to deer and rabbits. Flowers attract many pollinators.

Marsh marigolds serve as a welcoming sight for Hymenoptera in the early spring. These brightly coloured flowers offer a vital source of nectar and pollen to Hymenoptera after a long winter. As Hymenoptera forage for these resources, they play a crucial role in the ecosystem by inadvertently transferring pollen between flowers, promoting successful pollination for the marsh marigold. In this mutually beneficial relationship, the marsh marigold nourishes Hymenoptera while ensuring its reproduction.

Purple Flowering Raspberry
Rubus odoratus

◇◇◇◇◇◇◇◇◇◇◇◇◇◇◇◇

Growing Habits:

Mature Height: 6 foot

Ontario Hardiness Zone: 3 to 7

Part Shade to Full Sun

Medium

Sand Loam, Loam

Rubus odoratus is a strong-growing, deciduous shrub forming a thicket of erect stems clothed with palmate, 5-lobed, maple-like, dark green leaves, becoming pale yellow in the fall. The stems are thornless, unlike many other rubus species. From early to late summer, a ravishing display of large, fragrant, rich pink-purple flowers, 2 in. across (5 cm), can be enjoyed. Resembling single roses, they are borne singly or in few-flowered clusters above the handsome foliage. They are followed by small, fuzzy, purplish- red raspberries, which are edible, but tend to be insipid. However, they are valuable seasonal food for songbirds, game birds, or large and small mammals. Relatively immune to pests and diseases, Flowering Raspberry spreads rapidly from creeping underground stems and can form large colonies. A good plant for natural area or wildflower garden

The vibrant Purple flowering raspberry is a haven for Hymenoptera. These pollinators flock to the plant for its irresistible offerings: abundant nectar, easily accessible blooms, and a long blooming season. This mutually beneficial relationship thrives as the raspberry nourishes the Hymenoptera, and the Hymenoptera, in turn, aids in the plant's pollination.

By incorporating these native plants into your garden, you're creating a haven for a diverse array of Hymenoptera. These essential pollinators and predators are vital for maintaining a healthy ecosystem. Remember, providing a continuous supply of nectar and pollen throughout the growing season is crucial for supporting these beneficial insects. With careful planning, your garden can become a buzzing oasis for bees, wasps, and other fascinating Hymenoptera.

Lepidoptera

The Enchanting Lepidoptera in Ontario Gardens

◇◇◇◇◇◇◇◇◇◇◇◇◇◇

Ontario's diverse landscapes, from the dense forests of Algonquin Park to the serene meadows near Lake Huron, are graced by a unique array of Lepidoptera. These ethereal creatures, moths and butterflies, bring magic to the province's varied ecosystems. Imagine yourself in a wildflower-studded meadow, where the air is alive with the gentle flutter of butterfly wings and the delicate dance of nocturnal moths beneath the moonlit canopy.

What sets these enchanting insects apart? Picture the intricate scales embellishing the wings of the Eastern comma (*Polygonia comma*), each a miniature masterpiece, weaving patterns and colours that border on the otherworldly. Amidst this natural marvel, it becomes apparent that Lepidoptera are not simply insects but living canvases adorning the landscape with ephemeral beauty. The diverse habitats of Ontario provide a sanctuary for many Lepidoptera species, each adapted to specific niches, contributing to the region's rich biodiversity.

As we embark on a journey to uncover the mysteries of these winged marvels, we find that the realms of science and poetry intertwine in their extraordinary lives. The captivating world of Lepidoptera is not just about their beauty but also their vital roles in pollination. Their intricate life cycles, from egg to caterpillar to chrysalis for butterflies and cocoons for moths and finally to the emergence of a winged adult, add a layer of intrigue to their existence. Beyond their aesthetic allure, Lepidoptera serve as bioindicators, mirroring the health of ecosystems and responding to environmental changes. It's a world that sparks our curiosity, beckoning us to delve deeper into the wonders of nature.

By exploring the interconnectedness of these creatures with their surroundings, we aim to unveil the unknown that makes Lepidoptera critical to Ontario's ecology. This journey enhances our understanding of the interrelationship between nature's beauty and scientific wonder, inviting gardeners to admire and conserve the delicate balance that these winged wonders contribute to the fascinating landscapes of Ontario. It's a call to action, a reminder of our responsibility to protect and preserve these delicate creatures and their habitats.

Hickory Tussock Moth (Larva),
Lophocampa caryae

Top Left: Great Spangled Fritillary, *Argynnis cybele*

Top Right: Cecropia Moth, *Hyalophora cecropia*

Bottom: Viceroy, *Limenitis archippus*

What are Butterflies and Moths?

◇◇◇◇◇◇◇◇◇◇◇◇◇◇

The fascinating world of Lepidoptera, or butterflies and moths, is filled with wonder and enchantment. The very name "lepidoptera" originates from the Greek words "lepis," which means scale, and "pteron," which means wing. These remarkable creatures are adorned with intricate scales forming mesmerizing patterns on their wings, resembling a mosaic of iridescent tiles.

The life cycle of Lepidoptera is a wondrous journey through four distinct stages: egg, larva (caterpillar), pupa (chrysalis or cocoon), and adult. The journey begins with laying eggs on host plants, from which voracious larvae hatch and feed tirelessly. As they grow, the larvae enter the pupal stage, undergoing remarkable transformative changes. Finally, the adult butterfly or moth emerges from the pupa, showcasing a stunning evolution in form and function.

Discerning the differences between moths and butterflies unveils a world of intrigue. Butterflies, the elegant residents of the day, proudly display their vibrant colours in the warm sunlight, their club-shaped antennae guiding them gracefully from one flower to another. In contrast, moths, the enigmatic explorers of the night, reveal themselves in the serene twilight, their feathery antennae finely tuned to the subtle scents of the moonlit darkness.

Understanding the life cycle of Lepidoptera adds depth to their allure. It's a metamorphic journey akin to a captivating tale of transformation. Imagine a tiny egg delicately suspended on a leaf, hatching into a voracious caterpillar that devours foliage until finding a secluded spot to undergo a magical metamorphosis. Envision the chrysalis or cocoon as a mysterious capsule of possibilities, where the caterpillar undergoes a profound transformation, emerging as a glorious adult butterfly or moth. Witnessing this metamorphosis is like seeing the birth of living art, a testament to the wonders of nature.

Lepidoptera also holds a vital role as pollinators, significantly contributing to plant reproduction. While not all species are avid pollinators, many butterflies and moths actively transfer pollen between flowers as they seek nectar. Imagine a butterfly gently alighting on a flower, its proboscis probing for nectar and inadvertently carrying the essence of one bloom to another, facilitating the crucial pollination process. This process, often overlooked, is a cornerstone of biodiversity, highlighting the essential role of Lepidoptera in our ecosystems.

With their scale-covered wings, Lepidoptera embody a story of evolution, adaptation, and the delicate interplay between flora and fauna. As we explore the hidden world of these winged marvels, let us envision them not just as insects but as ambassadors of nature's beauty and biodiversity.

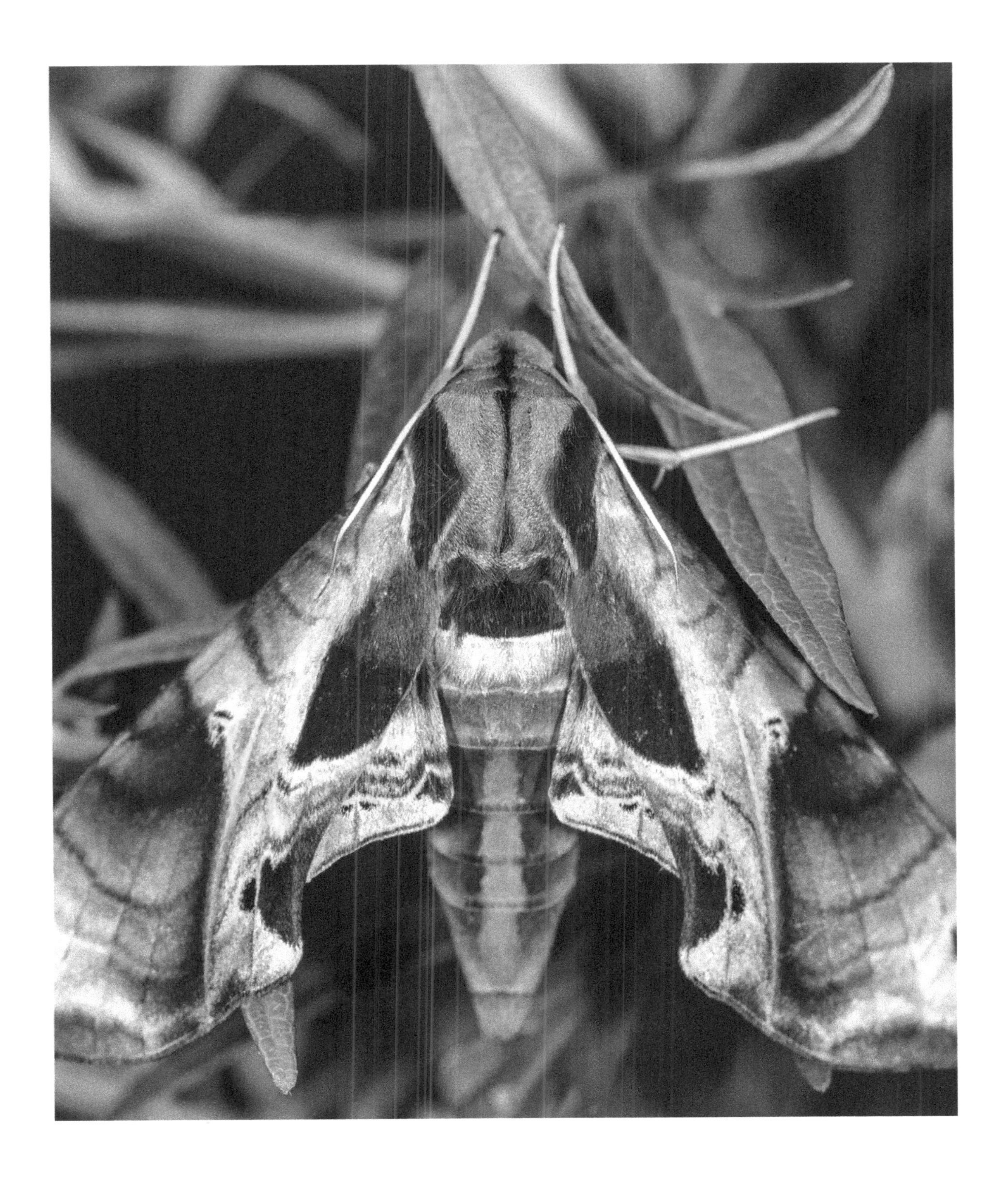

Pandorus Sphinx,
Eumorpha pandorus

The Importance of Lepidoptera in Gardens

◇◇◇◇◇◇◇◇◇◇◇◇◇◇◇◇

Pollination Partners

With their graceful flight and vibrant colours, Lepidoptera plays a crucial role as pollinators in our gardens. Their beauty is a visual delight and a testament to their essential role in the garden's vibrant tapestry. Envision the subtle transfer of life-sustaining pollen orchestrated by the movements of these winged wonders, and you'll truly appreciate their ecological significance.

Biodiversity Guardians

The presence of Lepidoptera in our gardens enhances biodiversity, creating a dynamic ecosystem. Picture a garden teeming with life—butterflies sipping nectar from blossoms, moths blending seamlessly into the moonlit foliage. In this biodiverse haven, Lepidoptera plays a crucial role in maintaining the delicate balance of species that coexist in harmony. By welcoming them, we contribute to the health and resilience of our ecosystems.

Caterpillars as Caretakers

In their caterpillar stage, Lepidoptera establish vital connections with specific host plants. While these voracious feeders consume foliage, their selective preferences contribute to the health and resilience of certain plant species. Picture a caterpillar nibbling on leaves, fostering a coevolutionary dance that ensures the survival of both insect and plant. In this nuanced relationship, Lepidoptera acts as caretakers, nurturing the vitality of their chosen botanical partners.

Environmental Barometers

The presence or absence of specific Lepidoptera species serves as a nuanced barometer of environmental health. Monitoring fluctuations in butterfly and moth populations provides insights into the overall well-being of ecosystems. Imagine a landscape where researchers meticulously track the population dynamics of a rare butterfly or moth, unravelling clues about habitat conditions and potential conservation needs, such as we do for Swallowtail (*Family Papilionidae*) butterflies. In this way, Lepidoptera becomes an indicator, subtly signalling the ecological nuances of their surroundings.

Educational Ambassadors

Beyond their ecological roles, Lepidoptera serve as more than just ambassadors of nature in educational contexts. They're captivating teachers. Gardens focused on these winged wonders offer immersive learning experiences. Picture a child observing the transformation of a caterpillar into a butterfly, eyes widening with surprise. In cultivating an appreciation for the intricacies of life cycles, metamorphosis, and interconnected ecosystems, Lepidoptera become more than just catalysts for environmental education. They become the heart of it. In this sense, gardens become more than just living classrooms; they become the gateway to a deeper understanding of and connection with the natural world.

In the interplay between Lepidoptera and our gardens, these enchanting insects emerge not just as visitors but as vital contributors to the flourishing tapestry of life. Their significance extends beyond the visible beauty they bring. They weave into the very fabric of our ecosystems, ensuring resilience, diversity, and the perpetuation of life.

Snowberry Clearwing,
Hemaris diffinis

Morphological Marvels
Anatomy of Lepidoptera

◇◇◇◇◇◇◇◇◇◇◇◇◇◇

The most recognizable characteristic of Lepidoptera is its wings, which are adorned with scales and allow it to exhibit a rich array of colours and patterns. Generally, pigmented scales with detailed designs enable identification, regulate heat energy, and provide aerodynamics during flight. This fragility and durability of the wings point out how precisely Lepidoptera has evolved to fit different environmental niches.

The adult mouthparts of lepidopterans, including those modified for nectar feeding found in butterflies or others adapted for diverse feeding strategies among moths, indicate their food preferences and ecological roles. Interestingly enough, many adult moths do not have functional mouthparts at all, as they only mate and then lay eggs, which in turn will eventually turn into pupae. Conversely, the caterpillar stage is characterized by chewing mouthparts, which enable these insatiable feeding larvae to consume vast quantities of plant materials to sustain their rapid growth and development.

The metamorphic life cycle of Lepidoptera adds another layer of wonder to their anatomy. Their change from egg to larva (caterpillar), pupa (chrysalis or cocoon), and finally culminating in adulthood reveals the ingeniousness behind the adaptability shown by these insects. The intricate process allows them to colonize diverse habitats and exploit various food resources, thus contributing significantly to their ecological success.

Lepidoptera uses sensory organs such as compound eyes and antennae for survival purposes, especially regarding reproduction. Large compound eyes' wide-range visibility is essential for navigation and mating. In contrast, antennae with chemoreceptors are also designed to detect pheromones and other chemical cues that aid communication and habitat selection.

The anatomy of Lepidoptera is a masterwork of nature's inventiveness, fusing delicate beauty with necessary adaptations. From scaled wings and modified mouthparts to a metamorphic life cycle and elaborate sensory organs, butterflies and moths add to gardens' ecological diversity and attractiveness. For gardeners, understanding the morphological marvels of Lepidoptera provides insights into creating spaces that support the vital roles of these enchanting insects. By growing gardens that acknowledge the ecological significance of Lepidoptera, we contribute to habitat conservation and ensure that they remain part of our natural environment's beauty.

Monarch butterfly,
Danaus plexippus

Danaus plexippus life cycle, egg, larvae, pupa (chrysalis), adult.

Positives and Negatives
Welcoming Lepidoptera in our Gardens

◇◇◇◇◇◇◇◇◇◇◇◇◇◇

Gardens are transformed by the beauty and mystery of Lepidoptera, a group of insects that includes butterflies and moths for their bright colours and remarkable patterns. Knowing the positive and negative effects of Lepidoptera infestation in your garden is essential when raising a healthy, balanced system.

One of the significant benefits of hosting Lepidoptera in a garden is their role as essential pollinators. Butterflies are excellent pollinators because they can visit a wide variety of flowers to drink nectar and spread the pollen across many species of flowers. Their long proboscis helps them to get into deep parts of flowers, enabling fast pollen transfer and promoting reproduction in flowering plants. In addition, some moth species contribute to pollination; some are known to be better pollinators than their day-flying counterparts, butterflies. As an Order, Lepidoptera extensively contributes to garden biodiversity and our ecosystems.

In addition, Lepidoptera are an essential source of nourishment for numerous other inhabitants within the ecosystem. Caterpillars are the larval form of butterflies and moths, which play an integral part in the food web as places where birds, spiders, and other predatory insects come to eat. As such, promoting Lepidoptera in your garden supports a multifaceted and interrelated community of organisms.

However, some challenges regarding possible limitations are based on particular Lepidoptera. Some species of caterpillars, such as Tomato hornworms (*Manduca quinquemaculata*), may eat voraciously and damage specific vegetation. These situations can also be managed effectively through regular monitoring and early intervention so that the damage they cause to vegetation is minimal.

Further, butterflies and moths can add aesthetic value to the garden, but some species may lay their eggs on specific host plants where localized defoliation occurs. Knowing the life cycles and tastes of various Lepidoptera species can help gardeners plan where they plant host plants, away from the most expensive ornamental or edible crops in gardens.

Embracing Lepidoptera in your garden leads to the beautification and biodiversity of the ecosystem. They perform a role in pollination and food for other wildlife, such as birds and bats, as this can increase the garden's overall health. Misguided challenges surround certain caterpillar species that can devastate some plants. Still, gardeners can easily manage these through informed planting and monitoring practices that help them achieve balanced and rich ecological growth. Pesticides should only be applied if they are necessary to maintain plant health. Using preventive cover sprays, where pesticides are sprayed several times a year on a calendar basis, has been shown to create more pest problems than it solves. Cover sprays generate the potential for pesticide runoff, increased human and pet exposure, and pest problems by suppressing predators, parasitoids and diseases that keep plant pests under control. Safer options could be to physically pick off invasive caterpillars and place them into a bucket of soapy water to drown them, insecticidal soaps when sprayed at dawn or dusk, and horticultural oils.

Mourning Cloak,
Nymphalis antiopa

Middle: Peck's Skipper,
Polites peckius

Bottom: Rosy Maple Moth,
Dryocampa rubicunda

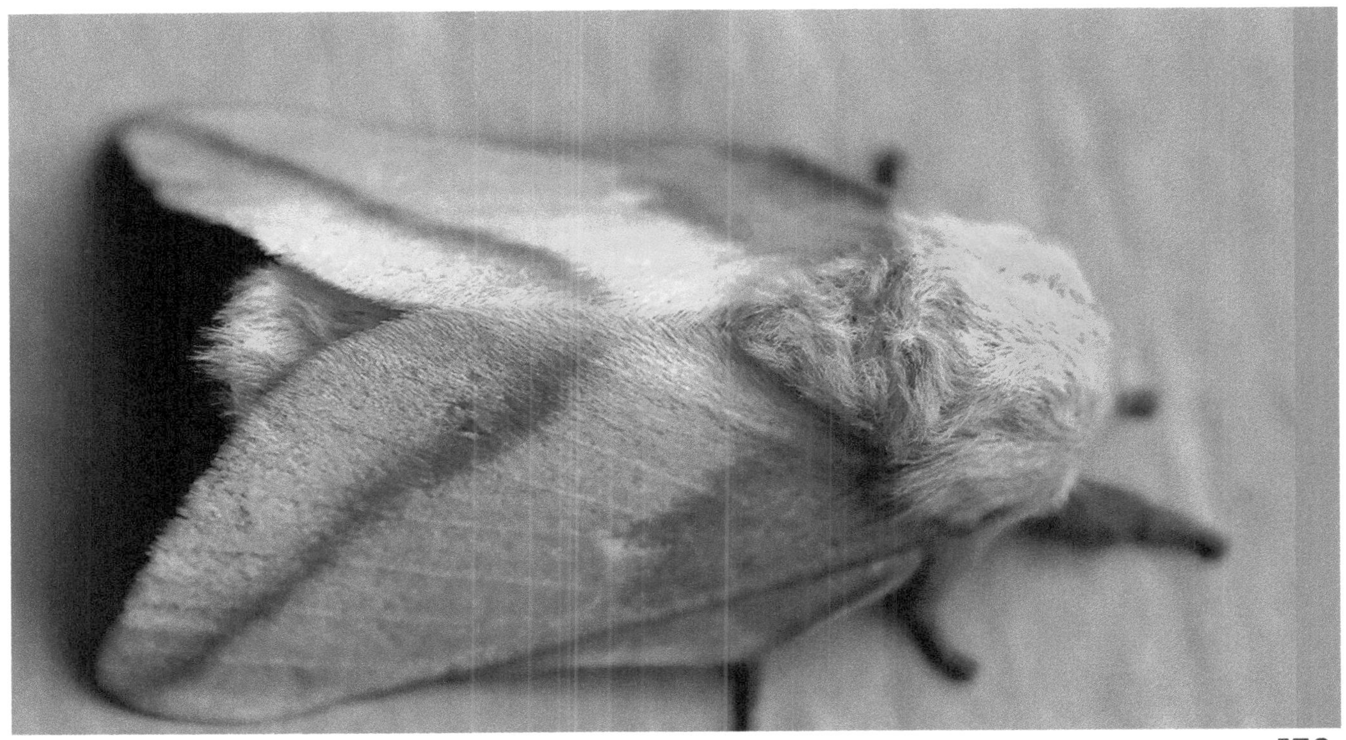

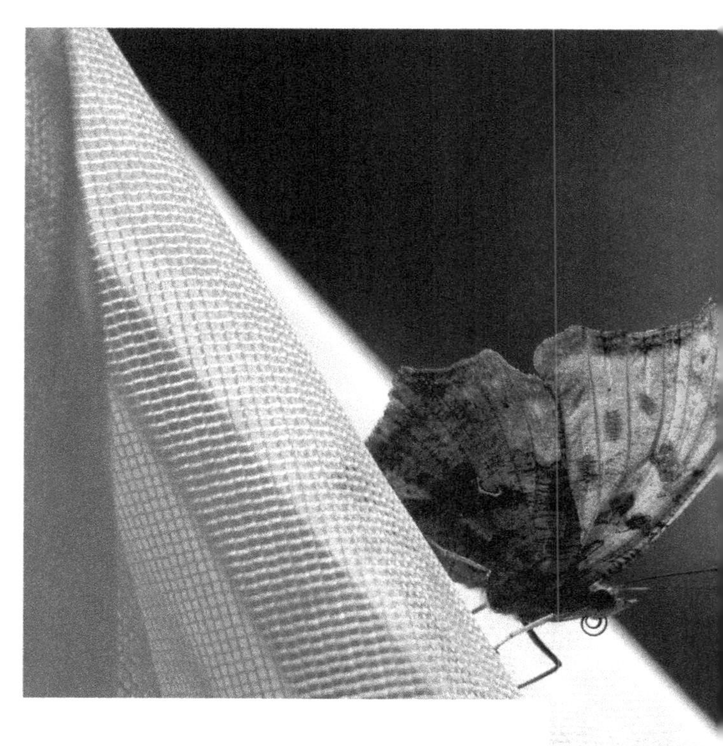

Discover Lepidoptera

Ontario's Rich Lepidoptera Diversity

Ontario is a treasure trove of Lepidoptera, with over 3000 species making their home here. Moths, in particular, have a strong presence, their numbers impressing even the most seasoned scientists. Each species, from the iconic Monarch butterfly (*Danaus plexippus*) to the subtly patterned Luna moth (*Actias luna*), plays a crucial role in maintaining the delicate balance of Ontario's ecosystems. Let's look at some common Lepidoptera that grace Ontario's gardens.

Here Is a Closer Look At The Top Five Lepidoptera In Ontario Gardens

Monarch Butterfly
Danaus plexippus

◇◇◇◇◇◇◇◇◇◇◇◇◇◇◇◇

Size: Adult range in size between 89 - 102 millimetres.

Description Tips: Monarch butterflies are a sight to behold, their vibrant burnt orange wings adorned with intricate black veins and delicate white spotted margins. The males' distinctive dark scent pouch on their hindwings is unique. Equally captivating are the caterpillars, with their vivid yellow, black, and white stripes. They grow to about two inches before undergoing a mesmerizing metamorphosis. In a fascinating display of behaviour, the females carefully lay a single egg on a milkweed plant, usually on the underside of a leaf near the plant's apex. These tiny eggs, about the size of a pinhead or pencil tip, are off-white or yellow and are adorned with longitudinal ridges running from the tip to the base.

Danaus plexippus, is a well-recognized and admired species in Ontario and across North America, celebrated for its extraordinary characteristics and behaviours. One of the most remarkable feats of the monarch butterfly is its annual long-distance migration, spanning over 3,000 miles. This awe-inspiring journey takes place as monarchs travel from their wintering grounds in the mountains of central Mexico to Ontario in early summer, and then, in the fall, millions of adult monarch butterflies east of the Rocky Mountains migrate to Mexico, while those in the western provinces of Canada migrate to various locations along the coast of California.

During their caterpillar stage, monarch butterflies have a highly specialized diet, feeding exclusively on native milkweed plant species. This unique diet sustains them and allows them to assimilate a chemical toxin known as cardenolides, which is toxic to birds and other potential predators. The vibrant burnt orange and black coloration of monarchs serves as a visual warning to predators, a remarkable adaptation that helps deter threats and ensures their survival. By planting native milkweed, we can play a crucial role in supporting these beautiful creatures.

Virginia Ctenucha Moth
Ctenucha virginica

◇◇◇◇◇◇◇◇◇◇◇◇◇◇◇

Size: Adult range in size between 40-50 millimetres.

Description Tips: The Virginia Ctenucha moth, North America's largest and most broad-winged wasp moth, is a sight to behold. Its metallic blue body, set off by a bright orange head and collar, is a striking contrast. The fore wing, a deep grayish brown with a hint of metallic blue at the base, and the black hind-wing, complete its unique appearance. This unique beauty is a marvel of nature, captivating all who have the privilege to see it.

Two generations of this moth grace us each year, a testament to the beauty of nature. The adults, with their metallic blue bodies and bright orange collars, are a sight to behold as they flutter among the flowers. Their role as essential pollinators, transferring pollen between flowers as they feed on nectar, is a crucial part of the ecosystem. This process is vital to the reproduction of many plant species, a fact that we should all be grateful for.

But the Virginia Ctenucha is not just a pretty sight. Its striking appearance serves a purpose beyond aesthetics. The moth's metallic blue body and bright orange collar are not just for show-they mimic the more dangerous wasp, deterring potential predators. This survival strategy is a fascinating aspect of the moth's life, making it more than just a beautiful creature. The adult Virginia Ctenucha flies primarily during the day but may also come to light at night. The larva body surface is black, covered with tufts of cream-coloured or black hairs. Caterpillar hosts include grasses, sedges, and irises.

Cabbage White Butterfly
Pieris rapae

◇◇◇◇◇◇◇◇◇◇◇◇◇◇◇◇

Size: Adult range in size between 32- 48 millimetres.

Description Tips: The Cabbage white butterfly, known as the *Pieris rapae*, is a genuinely mesmerizing insect. It boasts dazzling, pristine white wings adorned with a bold, striking black tip. You'll find a single spot on their wings in males, while females sport two on their forewings. However, the real surprise comes when you peek at the underside of their hindwings, revealing a unique unmarked greenish-yellow hue that is simply breathtaking. As for the larvae, they are velvety green with a delicate, faint yellow dorsal stripe and a row of charming yellow spots that elegantly run along their bodies. The oblong eggs are initially a striking white to cream colour, gradually maturing into a beautiful yellowish hue. These remarkable eggs are carefully laid on the underside of leaves, boasting a shape often likened to that of a football.

Pieris rapae, was a European species unintentionally introduced to North America in Quebec, Canada, in 1860. Since then, this species has adapted and thrived, becoming widely distributed across most of North America. It is particularly conspicuous in vegetable gardens and agricultural fields due to the abundant availability of larval host plants within the mustard family, including but not limited to cabbage, broccoli, and turnips. This adaptability is a testament to the resilience of Lepidoptera.

Cabbage white butterflies thrive in temperate climates and can be seen in a variety of habitats, including meadows, fields, gardens, and even open woodlands. Their adaptability is partly due to their short life cycle. Adults only live for about two weeks, but during that time, they can lay hundreds of eggs on plants in the Brassicaceae family, which includes cabbages, broccolis, and kales. The larvae, known as cabbage worms, are green and velvety with a distinctive yellow stripe running down their backs. These hungry caterpillars devour the leaves of their host plants, making them a pest for farmers who cultivate brassica vegetables.

Isabella Tiger Moth
Pyrrharctia isabella

◇◇◇◇◇◇◇◇◇◇◇◇◇◇◇

Size: Adult range in size between 40-53 millimetres.

Description Tips: *Pyrrharctia isabella* is a striking insect with wings that exhibit a beautiful spectrum of colours, ranging from vibrant orange-yellow to a warm, yellowish brown. Upon closer inspection, one can discern faint brownish antemedial, median, and postmedial lines adorn the pointed forewings, adding a delicate intricacy to the moth's appearance. Notably, the hindwings of the male Isabella tiger moth are a captivating shade of orange, while those of the female boast a more subdued yet charming rose-coloured hue, often adorned with subtly obscured spots.

Both male and female Isabella tiger moths possess abdomens adorned with rows of distinctive spots, contributing to their overall visual allure. In their larval stage, these moths are called "banded woolly bears" or simply "woolly bears." The caterpillars are densely covered in a soft, fuzzy texture, featuring prominent bands of black at both ends and a striking reddish-brown coloration in the middle, adding to their distinctive and memorable appearance.

Pyrrharctia isabella caterpillars, known as "woolly bears" due to their fuzzy appearance, have a captivating and intricate life cycle. In a manner reminiscent of bears, they enter a hibernation-like state during the winter, braving freezing temperatures as fully-grown caterpillars. Before hibernation, the larvae produce elevated glycerol levels in their blood, serving as a natural antifreeze that enables them to withstand the harsh winters in Ontario. *Pyrrharctia isabella* caterpillars pupate within cocoons constructed from their hairs, emerging as adults in the spring. There are two generations of banded woolly bear caterpillars each year, one in May and the other in August. The second generation, observed during late fall, hurries to find a place to spend the winter under bark or inside rocks or logs. In early spring, the woolly bear emerges from its winter rest as the days lengthen and the sun warms the earth. The caterpillar will briefly feed before seeking a safe and secluded location to transform into an adult. During this period, it forms a cocoon by producing silk and blending it with body hair. By late May or early June, the cocoon will begin to wiggle and then split open for the new moth to emerge as the year's first generation.

Red-spotted Admiral
Limenitis arthemis

◇◇◇◇◇◇◇◇◇◇◇◇◇◇◇◇

Size: Adult range in size between 75 - 100 millimetres.

Description Tips: *Limenitis arthemis*, also known as the red-spotted purple or white admiral, is a beautiful butterfly species native to North America. It's part of the Nymphalidae Family, commonly referred to as brush-footed butterflies due to their characteristic curled-up forelegs that resemble brushes.

The mature *Limenitis arthemis* butterfly boasts an upper side that ranges from blue to blue-green, with a captivating iridescence on the outer part of the hindwing. On the underside, the butterfly exhibits a striking dark brown coloration. Notably, the forewing is adorned with two eye-catching red-orange bars near the base, while the hindwing features three red-orange spots near the base, along with a row of additional red-orange spots. As for the caterpillars, they bear a remarkable resemblance to bird droppings throughout all stages of their development. These caterpillars may present as mottled brown or green with creamy blotches and are characterized by two knobby horns on the thorax.

When partially grown, caterpillars from the third brood enter winter dormancy within a specially rolled leaf known as a hibernaculum, which they silk to a branch. Their feeding and development then resume in the subsequent spring. The caterpillars undergo several moults throughout their growth cycle, shedding their old skin as they grow. Additionally, they undergo colour and pattern changes as they mature. To differentiate Red-spotted purple caterpillars (*Limenitis Archippus*) from Viceroy caterpillars, one can look for rounded rather than spiked projections behind the head. When it comes to reproduction, the eggs of these butterflies are laid Individually on the tip of their host plant's leaf, exhibiting a green colour that transitions to gray as the embryo develops. Notably, the surface of the eggs is covered with hexagons adorned with small spikes

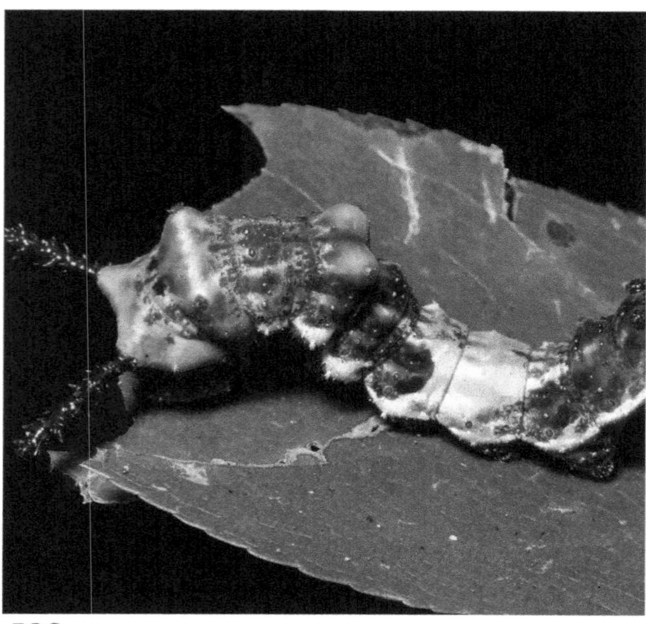

By creating a butterfly and moth-friendly garden, you're not only adding beauty but also supporting a vital part of the ecosystem. These delicate creatures play essential roles as pollinators and food sources for other wildlife. Remember, providing a variety of host and nectar plants is key to attracting and supporting different species. With patience and observation, you'll be amazed at the diversity of butterflies and moths that visit your garden.

Planting for Lepidoptera

Discover Native Ontario Plants

To create a suitable environment for Lepidoptera, we need to plant native flora that can serve as host plants for caterpillars and provide adult butterflies with a rich nectar source. Ontario Native Plants can help offer some native plants found in Ontario that can help attract and support the diverse world of Lepidoptera.

Cultivating a Lepidoptera-friendly garden not only adds aesthetic allure but also plays a crucial role in bolstering Ontario's biodiversity. By conscientiously blending visual appeal with ecological considerations, gardeners become active contributors to the mesmerizing choreography of butterflies and moths, enriching the local ecosystem and fostering a harmonious coexistence between nature and human cultivation. This symbiotic relationship not only enhances the liveliness of the garden but also underscores the vital role of sustainable practices in preserving and nurturing the diverse species that call Ontario home.

Mountain Mint
Pycnanthemum virginianum

◇◇◇◇◇◇◇◇◇◇◇◇◇◇◇◇

Growing Habits:

Mature Height: 3 feet
Mature Spread: 2 feet

Ontario Hardiness Zone: 3 to 7

Part Shade to Full Sun

Dry to Medium

Sand Loam, Loam, Clay Loam

Pycnanthemum virginianum, also known as common mountain mint or Virginia mountain mint, is a tall and sturdy herbaceous perennial plant that belongs to the mint family, Lamiaceae. This plant can reach up to 3-4 feet in height and is characterized by its square stems and lance-shaped, toothed leaves. During the summer and fall, *Pycnanthemum virginianum* is adorned with dense clusters of tiny, tubular, white flowers that are highly attractive to pollinators, especially bees and butterflies. The minty fragrance of the foliage further adds to its appeal in the garden.

This plant thrives in various growing conditions, including moist to slightly dry soils, and can tolerate partial shade. However, it generally performs best in moist, well-drained soils. While it is relatively low-maintenance, adequate moisture, especially during dry periods, is vital to encourage lush growth and prolific flowering. Whether grown in a garden border, wildflower meadow, or naturalized planting, Virginia mountain mint is a valuable addition to any landscape, offering beauty, fragrance, and ecological benefits.

Mountain mint, a delightful haven for Lepidoptera, is irresistibly attractive to these pollinators. They are drawn to this plant for abundant nectar, a sweet fragrance, and a long blooming season. The clusters of tiny flowers brim with nectar, providing a sugary fuel source for Lepidoptera throughout their active periods. The fragrant blooms emit a sweet scent that butterflies and moths can detect from afar, effectively guiding them to this delicious feast. Additionally, Mountain mints flower profusely throughout the summer, offering a reliable and extended source of sustenance for Lepidoptera throughout their breeding season. By providing consistent nourishment and attracting a diverse range of Lepidoptera, Mountain mint is vital in supporting healthy populations of these beautiful pollinators, sparking our curiosity about its benefits.

Common Milkweed

Asclepias silica

◇◇◇◇◇◇◇◇◇◇◇◇◇◇◇

Growing Habits:

Mature Height: 3 feet
Mature Spread: 2 feet

Ontario Hardiness Zone: 3 to 7

Part Shade to Full Sun

Dry to Medium

Sand Loam, Loam

This particular plant, known simply as milkweed, is a rapidly growing perennial that flourishes in a wide range of poor soil conditions, making it a versatile and resilient addition to any garden. Its significance in the Monarch butterfly's life cycle cannot be overstated, as it is a vital host plant for the species. Moreover, the plant's presence also attracts a diverse array of other butterfly species, adding to the overall biodiversity of the garden. It's common to observe caterpillars on the plant during the spring and early summer, as they utilize it as their sanctuary and food source.

This plant exhibits impressive resilience in drought conditions, making it a valuable asset in regions prone to water scarcity. However, its robust and vigorous growth habits are ideal for larger garden spaces with ample room to thrive. As an added delight, the plant graces the garden with its delicate, pale pink blooms during summer, adding a natural beauty to the surroundings. For those interested in exploring further options, other native species of Milkweed, such as Swamp milkweed (*A. incarnata*), Butterfly milkweed (*A. tuberosa*), and Poke Milkweed (*A. exaltata*), are also worth considering for their unique characteristics and contributions to the ecosystem.

Ironweed
Vernonia missurica

Growing Habits:

Mature Height: 5 feet
Mature Spread: 4 feet

Ontario Hardiness Zone: 4 to 7

Full Sun

Medium to Wet

Sand Loam, Loam, Clay Loam

Vernonia missurica, is a striking and resilient plant that adds a solid vertical presence to any garden. Its robust, iron-like stem gives it a sturdy and commanding appearance. In late summer, the plant bursts into intense and vivid purple blooms, creating a stunning focal point and attracting a diverse range of butterflies and other important pollinators to the garden.

This statuesque plant thrives in moist soils and can withstand short periods of flooding, but it also does well in average garden soils. To maintain a more compact and neat appearance, it's recommended to prune the stems in late spring. While Ironweed has the potential to spread aggressively through its seeds, this can be managed by removing some of the flower heads before they mature and disperse seeds.

Despite its potential for self-seeding, Ironweed's dark, sturdy stems adorned with feathery seed heads add an enchanting touch to the late-season landscape. It is an excellent choice for tall borders, cottage gardens, rain gardens, and wildflower meadows, infusing these landscapes with vibrant colour and visual interest.

Ironweed, with its vibrant flowers and nectar, is a beacon for butterflies and moths. These features, along with the blooms' pleasant fragrance, make it irresistible to Lepidoptera. But ironweed's role goes beyond mere attraction. It flowers throughout late summer and fall, providing a reliable and extended source of sustenance for Lepidoptera during their critical breeding season. By offering consistent nourishment and attracting a diverse range of lepidoptera, ironweed plays a vital role in supporting healthy populations of these beautiful pollinators. This is a testament to the power of nature and our role in preserving it.

Spotted Beebalm
Monarda punctata

◇◇◇◇◇◇◇◇◇◇◇◇◇◇◇

Growing Habits:

Mature Height: 2 feet
Mature Spread: 3 feet

Ontario Hardiness Zone: 3 to 7

Part Shade to Full Sun

Dry to Medium

Sand Loam

This particular plant, *Monarda punctata*, is well-known for its unique ability to attract a wide variety of pollinators with its beautiful clustered blooms. As a member of the mint family *Lamiaceae* and a native perennial plant to North America, it emits a delightful and sweet fragrance reminiscent of Greek oregano (*Origanum vulgare*). The tubular flowers of *Monarda punctata* are pale yellow with purple spots and appear on top of white to light purple bracts, creating a stunning visual display.

Not only appreciated for its ornamental value, this plant's dried leaves and flowers contain valuable medicinal properties traditionally used by various indigenous groups. Its leaves and flowers are known for their antifungal and antibacterial properties, making *Monarda punctata* a plant of interest in herbal medicine and natural remedies. *Monarda punctata* can be propagated by sowing seeds directly in the ground or pots and then transplanting them into sandy, well-drained soil. It can also be propagated by cutting young foliage, making it a versatile plant for cultivation. While it is tolerant of drought, watering in the summer can help keep the plants fresh and blooming for longer.

However, it's worth noting that this plant can become aggressive in certain growing conditions, spreading through rhizomes and self-seeding. Additionally, it is noticeably fragrant, making it a popular choice for at-tracting pollinators and adding a delightful aroma to gardens and natural landscapes.

Burr Oak
Quercus macrocarpa

◇◇◇◇◇◇◇◇◇◇◇◇◇◇◇◇

Growing Habits:

Mature Height: 80 feet

Ontario Hardiness Zone: 3 to 7

Full Sun

Dry to Medium

Sand Loam, Loam, Clay Loam

When selecting an oak tree for an urban setting, it's essential to consider the specific characteristics of the tree. Opt for slower-growing varieties, mainly those tolerant of heavy air pollution, to ensure the tree's resilience in an urban environment. Look for oak species that can adapt to various soil types, including clay, but preferably thrive in loamy soils with ample sunlight. Remember that these trees require significant space, as their roots are highly competitive. However, they develop into beautiful shade canopies with vibrant fall foliage when provided with sufficient room to grow.

In addition to their visual appeal, *Quercus macrocarpa* offer practical benefits. The acorns produced by these trees are not only edible and sweet but can also be ground into flour for making bread, adding a culinary aspect to the tree's appeal. Furthermore, oak species benefit local ecosystems by serving as host plants for over 500 Lepidoptera caterpillars, providing a critical habitat for these larvae. Oaks host more Lepidoptera caterpillars than any other tree genus, making them valuable to any urban environment.

Quercus macrocarpa is not just a tree. It's a nurturing habitat for many butterfly and moth species. These majestic trees, with their bur oak leaves, serve as a primary food source for many caterpillar species, offering a three-course meal for Lepidoptera. The Bur oak's dense canopy and rough bark provide a protective haven for overwintering pupae, a nurturing environment that ensures their survival. Moreover, female butterflies and moths choose these oak leaves to lay their eggs, ensuring a ready food source for their offspring. This nurturing role of the bur oak, providing sustenance, shelter, and breeding ground, makes it a critical player in supporting the entire life cycle of Lepidoptera populations.

By incorporating these native plants into your garden, you're providing essential food and habitat for a wide range of Lepidoptera. Remember, caterpillars rely on specific host plants for survival, while adult butterflies and moths require nectar sources. By creating a diverse plant community, you'll attract a variety of species and contribute to their conservation. Observe closely as you watch caterpillars transform into stunning butterflies and moths.

Neruoptera

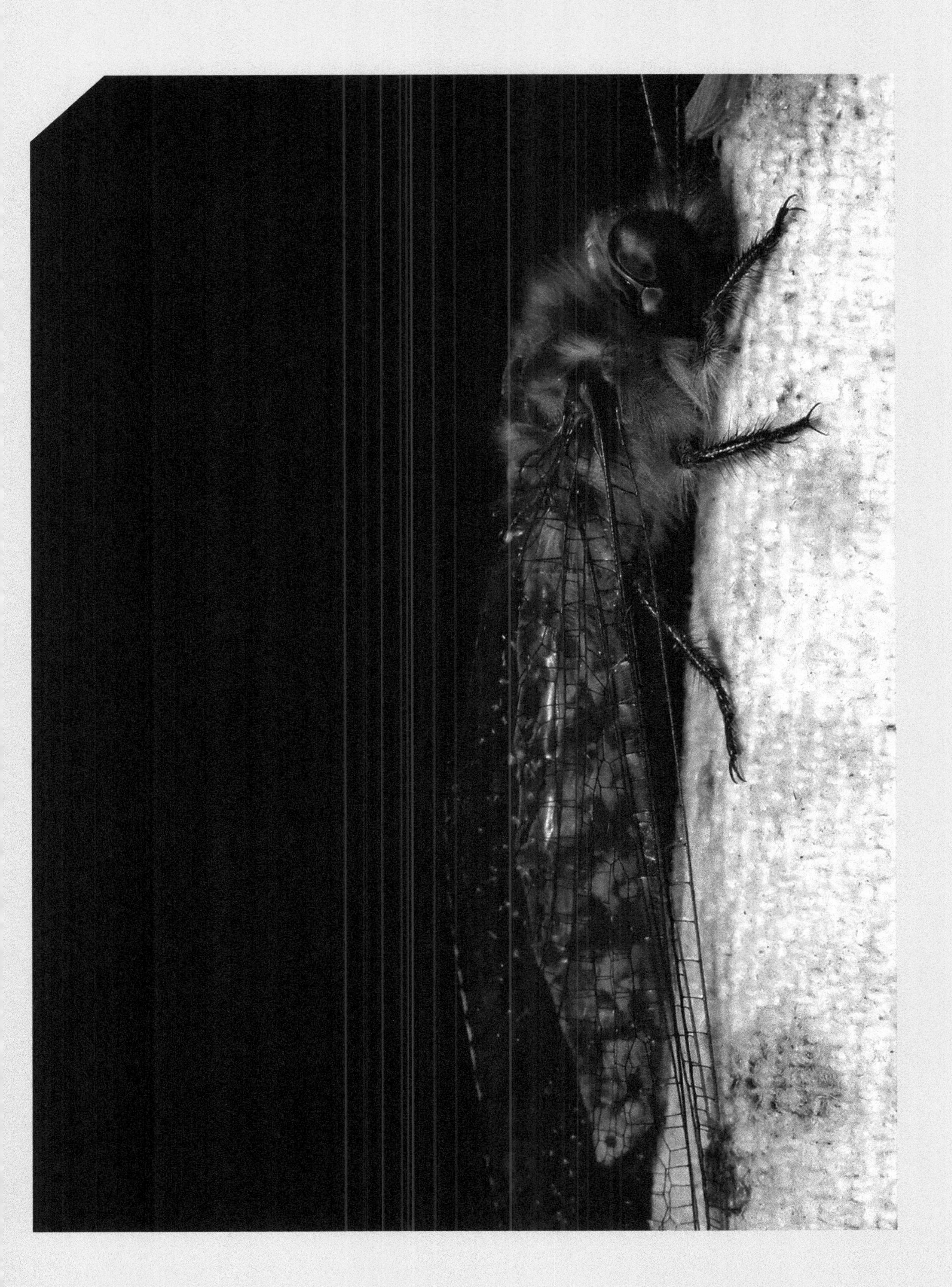

The Neuroptera of Ontario Gardens

◇◇◇◇◇◇◇◇◇◇◇◇◇◇

Explore the enchanting world of Neuroptera, a diverse order of insects that plays a vital yet often overlooked role in the lush gardens of Ontario. Among the vibrant flora, Neuroptera stands as a silent guardian, maintaining the delicate balance of nature. Picture a realm where lacewings, antlions, and owlflies orchestrate a captivating symphony of predation and pollination, intricately influencing your garden's ecosystem.

In the verdant landscapes of Ontario, over thirty species of Neuroptera take center stage, their significance extending beyond their intricate anatomy. Dive into the hidden world of these insects, where beauty is intricately woven into the delicate veins of their wings. Beneath their graceful appearance lies a realm of adept hunters, skillfully navigating the microcosm of small invertebrates.

Beyond their visual appeal, Neuroptera holds the key to ecological harmony in your garden. As natural pest controllers, they prey upon aphids and mites, showcasing a symbiotic relationship with the surrounding flora. Uncover the mysteries of Neuroptera and witness how their presence shapes the fabric of your garden, contributing to the resilience and balance of the natural world.

Let the foliage part to reveal the hidden gem, Neuroptera. As you unravel the layers of their existence, the gardens of Ontario will transform into dynamic ecosystems where every flutter of lacewing wings and every delicate trill of an owlfly contribute to the enchanting narrative of nature's delicate dance. Welcome to a world where the seemingly inconspicuous becomes a compelling revelation, and the pursuit of understanding unlocks the doors to a garden alive with the charm of Neuroptera.

Top Right: Say's Mantidfly, *Dicromantispa sayi*

Top Left: Red-lipped Green Lacewing (Larva), *Chrysoperla rufilabris*

Bottom: Eastern Spotted-winged Antlion, *Dendroleon obsoletus*

Masters of Aerial Elegance and Predation

◇◇◇◇◇◇◇◇◇◇◇◇◇◇

Neuroptera, an Order encompassing lacewings, antlions, and owlflies, presents an irresistible entomological sophistication. Distinguished by their meticulously crafted wings adorned with a mesmerizing net-like pattern of veins, these insects embody a unique blend of delicate beauty and predatory prowess. The name Neuroptera comes from the Greek "neuron," meaning nerve, and "pteron," meaning wing, which refers to the nerve-like network of veins on the wings of these insects.

Wings as Masterpieces of Evolution

The intricacy of Neuroptera wings is a testament to millions of years of evolutionary refinement. Composed of a network of finely interlaced veins, these structures serve as flight instruments and play a vital role in camouflage. The delicate patterns mimic foliage, allowing Neuroptera to seamlessly blend into their surroundings, a crucial adaptation for successful predation.

Predatory Excellence

Beneath the enchanting facade lies a world of formidable hunters. Neuroptera has honed predatory strategies that range from sophisticated hunting techniques to elaborate mimicry. Many species are voracious consumers of aphids, mites, and other garden pests, contributing significantly to biological pest control. Neuroptera's adaptability in various ecosystems underscores their role as key players in maintaining the delicate balance of biodiversity.

Specialized Adaptations

Within the Order Neuroptera, each subgroup—lacewings, antlions, and owlflies—displays distinctive adaptations for survival. For instance, green lacewings (*Family Chrysopidae*) utilize specialized mouthparts to puncture and consume prey. Antlions, with their iconic pit-trap larvae, patiently await unsuspecting prey to tumble into their carefully constructed sand traps. Meanwhile, owlflies showcase exceptional aerial agility, executing intricate maneuvers to capture prey mid-flight.

Chrysopa oculata

Life Cycle Marvels

Neuroptera undergoes a mesmerizing process of complete metamorphosis. The journey begins with some species strategically depositing eggs directly into sand or on vegetation. The eggs often have a distinctive pattern, each housing the potential for a remarkable transformation. Upon hatching, the larvae emerge as voracious predators, equipped with specialized mouthparts and formidable mandibles. As these larvae mature, they enter a pivotal pupal stage, encased in a protective cocoon or chrysalis. Internal restructuring occurs during this phase, shaping the intricate features of the adult Neuroptera.

Finally, the fully developed adult emerges, adorned with gossamer wings showcasing a net-like pattern of veins. This adult stage marks the culmination of a metamorphic journey as the Neuroptera takes to the skies, contributing to ecological balance through predation and engaging in the perpetuation of its species. The elegance of this metamorphic process underscores the adaptability and resilience of Neuroptera in navigating the complexities of their ecological niches.

Ecosystem Engineers

Far from mere predators, Neuroptera can also be considered ecosystem engineers. Their presence influences other insects' behaviour and population dynamics within a given habitat. Neuroptera indirectly fosters a more balanced and sustainable ecosystem by regulating pest populations, benefiting gardeners and the entire web of life.

Neuroptera's exquisite wings and predatory prowess elevate them to masterful aerial predators. As guardians of garden ecosystems, these insects showcase the intricate dance of evolution and the artistry of nature, weaving a story of survival, adaptation, and the unending quest for ecological equilibrium.

Lacewing eggs

Why Do We Need Neuroptera?

◇◇◇◇◇◇◇◇◇◇◇◇◇◇

Apart from being a natural pest control master within gardens, the Order Neuroptera claims a special place in every gardener's heart because of its contribution to natural pest control. Lacewing larvae, known for their strong jaws and voracious feeders, have an insatiable appetite for aphids, mites and other garden pests. This dynamic biological control system reduces reliance on chemical pesticides when lacewing (*Order Neuroptera*) larvae are released into the garden. By doing so, plant health is protected, and a balance between beneficial and harmful insects is maintained. They are ensuring environmental sustainability and resilience.

The elegance and precision that dominate the life cycle of Neuroptera are remarkable. Adult lacewings gently lay eggs on plant stalks singly or in batches. Some species will lay their eggs directly into sand or vegetation higher up plants. These strategically positioned eggs serve as a defence tactic that protects the upcoming generations of lacewings from potential predators. When gardeners understand and appreciate this aspect of lacewing reproduction, they can provide favourable conditions for these unique insects, perpetuating natural pest control and presence within the yard.

Neuroptera's usefulness goes beyond just being pest predators in gardens. Although lacewing larvae are popular because they are successful predators, adult lacewings, antlions, and owlflies, assist in pollinating flowering plants. Nectar attracts them, but pollen adheres to their body surfaces during feeding, making pollination possible. Neuroptera's adaptability is evident from this double purpose, which shows their importance in maintaining biodiversity and reproductive success among garden flora.

Having lacewings, antlions, and owlflies, as part of a gardener's toolkit proves how efficient and elegant Mother Nature's architects are. Lacewings are essential to garden ecosystems with their natural pest control capabilities, intricate life cycles, and adaptable feeding behaviour. By recognizing Neuroptera populations as significant contributors to their gardens' healthiness and stability, gardeners contribute to the health and resilience of cultivated spaces and ensure that a fragile balance is maintained to support myriad life forms. As custodians of these green sanctuaries, we should accept Neuroptera as the architects of beauty who create and improve our gardens through a dynamic interplay of predators and prey, life cycles, and the continuous regeneration of nature's design.

Humulin Brown Lacewing,
Hemerobius humulinus

Morphological Marvels

Anatomy of Neuroptera

◇◇◇◇◇◇◇◇◇◇◇◇◇◇

One feature that distinguishes Neuroptera from other insects is its complex wing venation. Their membranous wings resemble lacy patterns when seen from a distance, thus inspiring their common name, "lacewings." Due to their unique design for efficient flying, they make good examples for studying how evolutionary adaptation has led to survival in different habitats. Often used for taxonomic identification, venation highlights how vital this morphological feature can be when we examine how diverse this
Order is.

The mouthparts of Neuroptera reveal an exciting blend of adaptations for predation and liquid feeding. Several lacewings have long slender mouthparts suited for sucking nectar, while others, such as larvae, have strong jaws that can capture small prey. This diversity in feeding strategies places Neuroptera among the most valuable members in garden ecosystems, contributing to pollination and pest control services.

Neuroptera larvae have several adaptations specifically meant for predation. They create pitfall traps with elongated sickle-shaped mandibles (jaws) hidden inside well-camouflaged bodies to catch unsuspecting prey. The predatory success of Neuroptera's early stages makes them useful biological agents that help reduce pest numbers within the garden, thus contributing more to their ecological significance.

Neuroptera's reproductive strategies also show fascinating adaptations. Some species use elaborate courtship behaviours involving vibrational signals and pheromone exchange during mating. Some Neuroptera species use different reproductive strategies that can be grouped into sexual and asexual modes of reproduction.

An account of Neuroptera's anatomy becomes a story about the creativity of evolution. Information on Neuroptera's wing venation, mouthparts, and predatory adaptations among larvae is vital for understanding garden ecosystems' diversity and balance. For gardeners, understanding the morphological marvels of Neuroptera provides insights into fostering environments that support the critical roles of these beneficial insects. Cultivating gardens that consider the ecological importance of Neuroptera assists with maintaining resilient and sustainable landscapes where lacewings are viewed as valuable partners in promoting a healthy and vibrant ecosystem.

Red-lipped Green Lacewing,
Chrysoperla rufilabris

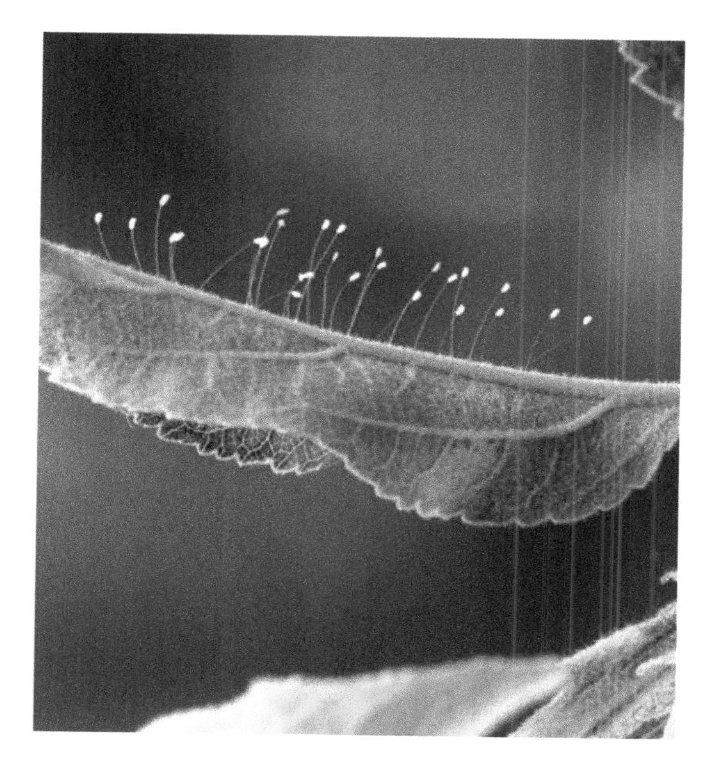

Chrysoperla rufilabris lifecycle egg, larva, pupa, and adult.

Positives and Negatives
Welcoming Neuroptera in our Gardens

◇◇◇◇◇◇◇◇◇◇◇◇◇◇◇◇

Lacewings are among the most intricate components of a garden ecosystem, with advantages and disadvantages for conscientious gardeners. The positive aspect of including lacewings, antlions, and owlflies in gardens is based on the ability of their larvae to act as natural enemies. Lacewing larvae, for example, not only have a great liking for common pests such as aphids, mites, and caterpillars but also favour pest eggs, thus helping in the complete control of pests. This preference for various life stages of pests suggests that lacewings can be a beacon of hope, an eco-friendly substitute for chemical pesticides, inspiring a more sustainable approach to pest management.

Adult lacewings are the most common species of Neuroptera found within Ontario gardens, and their multifunctional role goes beyond pest control measures. They pollinate by feeding on nectar and pollen, which assist in fertilizing flowers. According to findings from various research, lacewings (as well as antlions and owlflies) enhance the overall biodiversity of garden ecosystems through positive influences on plant-pollinator interactions by their predatory larvae.

However, closer examination reveals some challenges that gardeners may face. Although lacewings show some efficacy against pests, their impact may vary depending on the severity of pest infestations. If pests are under pressure, additional pest management methods may be required. In addition, some types of lacewing larvae feed indiscriminately; hence, they may consume beneficial insects inadvertently, resulting in disturbance within the web-like ecological relationships in a garden.

Nevertheless, gardeners' role in fostering biodiversity by incorporating lacewings, antlions, and owlflies into gardens is crucial. This aligns with sustainable and integrated pest management practices. Choosing native lacewing species adapted to local ecosystems can make them more effective biocontrol agents. Including diverse plant species providing suitable habitats and alternative food sources for lacewings would further maximize their presence within the garden, showcasing the power of your decisions in shaping the garden ecosystem.

Neuroptera plays a significant part in maintaining ecological balance within gardens. Their dual roles as efficient predators and pollinators emphasize their importance in sustainable pest management techniques. For example, gardeners can benefit from lacewings by considering variables such as local biodiversity, specific lacewing species, or the overall health of the garden ecosystem. Through informed and strategic management, lacewings become integral allies in fostering thriving and resilient garden environments.

Micromus posticus

Discover Neuroptera

Ontario's Rich Neuroptera Diversity

Ontario boasts a rich diversity of over forty Neuroptera species. From the delicate Green lacewings to the fierce antlions, enthusiasts can discover a fascinating array of these insects in various habitats. The Province's diverse ecosystems provide a home to numerous Neuroptera species, each adapted to specific environmental niches.

Here Is a Closer Look At The Top Five Neuroptera In Ontario Gardens

Golden-Eyed Lacewing
Chrysopa oculata

◇◇◇◇◇◇◇◇◇◇◇◇◇◇◇

Size: Adult range in size between 112-14 millimetres.

Description Tips: *Chrysopa oculata*, also known as the Golden-eyed lacewing, is a remarkable insect characterized by its elegant, elongated body. The thorax is embellished with two immaculate rows of tiny, obsidian-hued dots, while the head boasts a striking black-and-white motif, creating a visually arresting contrast. The most enchanting aspect of this creature is undoubtedly its resplendent, oversized eyes, which gleam with a lustrous gold or copper hue. Remarkably delicate and ethereal, its four gossamer wings are intricately veined with a mesmerizing network of verdant lines akin to a delicate lacework masterpiece. Adding to its allure, the insect's elongated, filamentous antennae exude an air of grace and sophistication.

Chrysopa oculata, is a captivating insect with a remarkable life cycle. This fascinating insect undergoes a complete metamorphosis, transitioning through two distinct life stages. As adults, they delicately feed on pollen, nectar, and aphid honeydew, making them essential pollinators and natural pest controllers. Their striking golden eyes and delicate wings make them a beautiful sight as they flutter around artificial lights in the evening. In contrast to their gentle adult counterparts, the larvae of the golden-eyed lacewing are fierce predators. These voracious hunters prey on various pests, including aphids, mites, leafhoppers, and even tiny caterpillars. Their predatory nature makes them valuable allies in natural pest control.

Found throughout North America, these nocturnal insects lay their eggs individually on foliage, with each female capable of laying several hundred eggs in her lifetime. The eggs hatch in three to six days, giving rise to voracious larvae that go through three stages over two to three weeks before pupating in a silk cocoon. The pupae overwinter in the soil, emerging as adults in the spring to continue the golden-eyed lacewing life cycle.

This intricate life cycle and the Golden-eyed lacewing's dual roles as a gentle pollinator and a fierce predator make it an essential part of the ecosystem.

Brown Wasp Mantidfly
Climaciella brunnea

◇◇◇◇◇◇◇◇◇◇◇◇◇◇◇◇

Size: Adults range in size between 23 -35 millimetres

Description Tips: *Climaciella brunnea*, commonly called the Brown wasp mantidfly, is a captivating insect with a remarkable and distinct appearance. Its slender body is adorned with eye-catching bands of vibrant yellow, deep brown, rustic hues, or intense black, strikingly resembling a fusion of a wasp and a praying mantis. What truly sets this insect apart is its exquisitely patterned and coloured wings, closely mirroring those of a wasp. This remarkable adaptation serves as a clever defence mechanism against potential predators. It's worth noting that the coloration of the brown wasp mantidfly varies significantly across its range in North America. This variability allows the insect to seamlessly adapt its appearance to blend in with the local population of wasps, presenting a captivating example of natural camouflage and adaptation.

The Brown wasp mantidfly is a fascinating insect commonly found in grasslands and open habitats. These remarkable creatures are active hunters, meticulously preying on various insects that visit flowers. Their hunting behaviour is intriguing, as they prefer certain insect species, showcasing their specialized predatory nature.

Climaciella brunnea larvae are parasitoids of spiders, adding a layer of complexity to its ecological role. This unique aspect of its life cycle highlights its impact on spider population dynamics, further emphasizing its significance in the ecosystem. Another fascinating behaviour of adult mantidflies is their enigmatic attraction to lights at night. This adds to their mystique and makes them a subject of interest for researchers studying insect behaviour.

When considering their role in natural ecosystems, it becomes evident that the impact of mantidflies on the populations of small invertebrates is not only individually significant but also plays a crucial part in maintaining the ecosystem's delicate balance. Their presence alongside predatory arthropods such as spiders, assassin bugs, robber flies, ants, lady beetles, and mantises contributes to the intricate web of interactions sustaining the ecosystem's stability. Despite their predatory nature, adult mantidflies are delicate and small, rendering them vulnerable to a wide variety of animals, including spiders, assassin bugs, birds, lizards, frogs, fish, shrews, and salamanders. This vulnerability underscores the complexity of their position within the food web. It emphasizes their importance as crucial links in the chain, ultimately contributing to the intricate balance of natural ecosystems.

Say's Mantidfly
Dicromantispa sayi

◇◇◇◇◇◇◇◇◇◇◇◇◇◇◇

Size: Adults range in size between 14 – 15 millimetres

Description Tips: *Dicromantispa sayi*, commonly known as Say's Mantidfly, is a fascinating insect with a striking appearance. It has a distinctive brown-to-black head adorned with pale yellow markings, including a longitudinal yellow stripe on each side of its face and a dark stripe down the center. The multifaceted eyes and mostly brown antennae, with the first segment being yellowish, add to its unique features.

Moving down to its elongated and slender neck (prothorax), the base and end of the neck are adorned with a cream-coloured patch, while the rest of the neck transitions from yellowish-brown to dark brown. The female's thorax sides are marked with dark, cream-coloured, curved lines, while the males are predominantly pale. The wings of the Say's Mantidfly are transparent with brown veins and lack brown spots at the wing tips. The base and outer edge of the wings (costa area) are dark brown, with a faint yellowish streak. The stigma at the end of the costa is also dark brown.

The front legs of the Say's Mantidfly are modified to resemble those of a praying mantis (*Order Mantodea*), dark brown with light streaks, and are typically folded up and not used, except for eating. The mid and hind legs are yellow with brown streaks, with female coxae brown and male coxae pale. The abdomen ranges from yellowish-brown to dark brown with yellow marks, with the female abdomen usually being darker and possibly entirely black, while the yellow marks vary considerably. The cream-coloured edges of the abdomen are more prominent on the male and more spotted on the female.

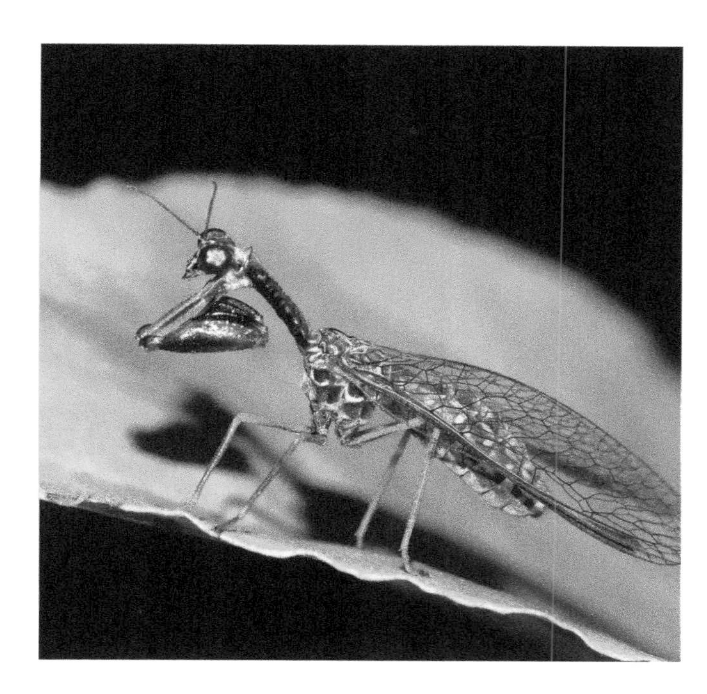

Four-spotted Owlfly
Ululodes quadripunctatus

◇◇◇◇◇◇◇◇◇◇◇◇◇◇

Size: Adults range in size between 30 – 38 millimetres

Description Tips: The Four-spotted owlfly, scientifically known as *Ululodes quadripunctatus*, is a fascinating insect species commonly found in Ontario, Canada. This remarkable insect boasts a truly unique appearance. It is distinguished by a sleek, black, damselfly-like body that measures approximately 1 1/2 inches (38 mm) in length. One of its most eye-catching features is its striking 1-inch (25 mm) knobbed antennae. The head and thorax of this extraordinary insect are adorned with short gray hairs, adding to its distinctive allure. The dorsal side of the Four-spotted owlfly's black abdomen is adorned with striking yellow markings, further enhancing its visual appeal.

In addition to its captivating physical attributes, the Four-spotted owlfly has predominantly transparent wings with darker stigmas and up to three distinct darker patches on the hind wings. This combination of colours and patterns makes for a striking display when the insect takes flight. One particularly intriguing behaviour the *Ululodes quadripunctatus* exhibits is its notable attraction to lights. This behaviour adds to the mystique surrounding this captivating insect and has piqued the interest of entomologists and nature enthusiasts alike.

Owlflies play a crucial role as predators in their ecosystems, significantly regulating insect populations. Throughout their life cycle, owlflies exhibit a range of fascinating survival strategies. For instance, owlfly larvae employ intricate camouflage techniques, mimicking twigs by adorning themselves with waste and debris. This clever adaptation helps them blend seamlessly into their surroundings, providing adequate concealment from potential predators. Furthermore, owlfly larvae engage in a behaviour known as "fencing" of trophic eggs, a remarkable tactic aimed at safeguarding themselves from predators. This behaviour underscores that, despite being formidable predators, owlflies are also vulnerable to becoming prey for other animals.

Immaculate Antlion

Myrmeleon immaculatus

◇◇◇◇◇◇◇◇◇◇◇◇◇◇◇

Size: Adults range in size between 30 millimetres

Description Tips: *Myrmeleon immaculatus,* undergoes a dramatic transformation throughout its life cycle. The larval stage, lasting up to three years, is where the action happens. These slender, camouflaged masters of disguise (up to 15mm long) reside in loose, dry sand, particularly in arid and semi-arid regions like sandy beaches, dunes, scrublands, or open woodlands. Their most impressive feat is constructing conical pits (2-5cm in diameter) using their bodies to meticulously fling sand out-wards. These pits become deadly traps for unsuspecting prey, primarily ants that lose their footing and tumble down the loose walls. The waiting larva at the bottom utilizes its signature weapon - a pair of long, sickle-shaped mandibles - to seize the fallen victim.

Here's where the true horror unfolds. The antlion injects digestive enzymes into its prey, liquefying the insides. Once the unfor-tunate victim becomes a nutritious soup, the larva sucks out the fluids, leaving behind an empty shell. However, the adult stage paints an entirely different picture. These winged insects, with a blue-grey body and a wingspan of up to 50mm, emerge for a short, dramatic performance. They are exclusively nocturnal and solely focused on finding a mate for reproduction. Adults lack functional mouthparts and don't feed, relying on energy reserves accumulated during their dominant larval stage. After a whirl-wind of mating activity, the female lays eggs in loose sand, perpetuating the cycle of the ingenious pit-dwelling predator.

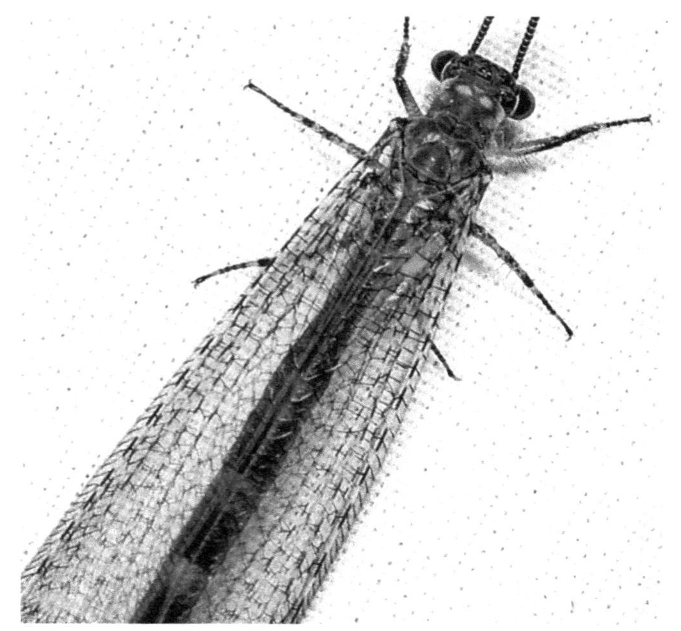

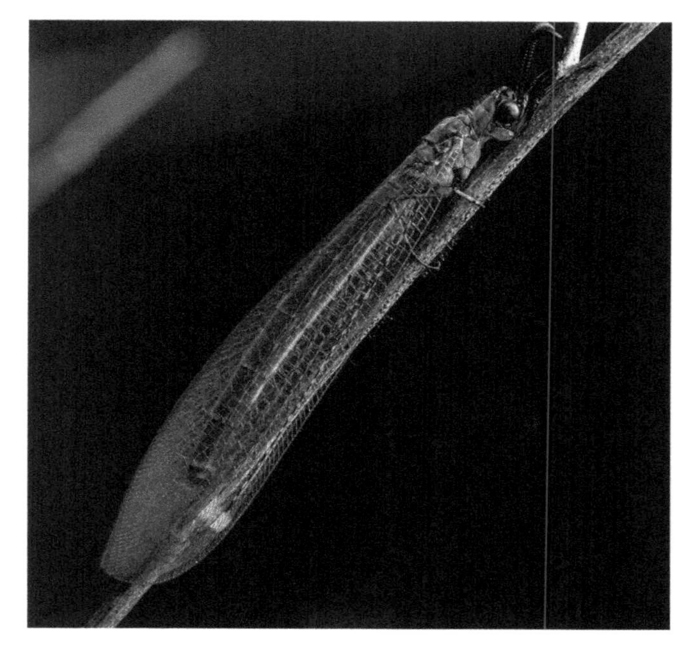

Neuroptera, often overlooked, are invaluable allies in the garden. Their predatory larvae and adult insects help to control populations of aphids, scale insects, and other pests. By creating a diverse garden with a variety of plants, you're providing suitable habitats for these beneficial insects. Remember, many Neuroptera have specific habitat requirements, so research the species in your area. By understanding their life cycles and roles in the ecosystem, you can create a garden that supports these fascinating creatures.

Planting for Neuroptera

Discover Native Ontario Plants

To create an irresistible haven for beetles in your garden, strategically planting native Ontario species goes beyond mere horticulture—it's an invitation to witness a thriving, interconnected ecosystem. Here's a closer look at some native Ontario plants from Ontario Native Plants that we suggest you, as gardeners, plant and their role in enticing and supporting the diverse world of beetles and other wildlife.

By strategically integrating these native Ontario plants and others into your garden, you're not just adding to its visual appeal but crafting a vigorous stage where beetles play pivotal roles. This carefully curated selection becomes an essential component of your garden's biodiversity, supporting beneficial beetle species and enhancing the overall ecological resilience of your outdoor space. As these plants bloom and flourish, they beckon beetles, turning your garden into a living tapestry where each species contributes to the intricate dance of life, from the ground-dwelling protectors to the aerial pollinators. Your actions are crucial in this delicate balance.

Harebell
Campanula rotundifolia

◇◇◇◇◇◇◇◇◇◇◇◇◇◇◇

Growing Habits:

Mature Height: 1 foot
Mature Spread: 1 foot

Ontario Hardiness Zone: 3 to 7

Part Shade to Full Sun

Dry to Medium

Sand Loam

The Harebell, scientifically known as *Campanula rotundifolia*, is a delicate and charming perennial plant found in various habitats across the Northern Hemisphere. Its slender and graceful stems grow in clusters, reaching heights of 4 to 15 inches. These fragile stems often cause the entire plant to gracefully arch and bend, adding to its unique and elegant appearance. The characteristics of the Harebell can display significant variation depending on the specific habitat conditions in which it grows.

The Harebell's basal leaves are rounded and tend to wilt early in the growing season, while the narrower stem leaves persist, adding to the plant's overall allure. The bell-shaped flowers of *Campanula rotundifolia* are a striking blue-violet colour and hang singly or in clusters along delicate, nodding, thread-like stems. These lovely, bell-shaped lavender flowers are borne in loose clusters at the tips of the stems, creating a beautiful visual display in their natural habitat.

The genus name, derived from the Latin word "campana," meaning "bell," aptly reflects the bell-like shape of the flowers. It's an appropriate homage to the plant's enchanting blossoms. Interestingly, the name "Harebell" has historical associations with witches, as there was a belief that they could transform themselves into hares. Encountering these plants was thought to bring bad luck, leading to another old name for the plant in Scotland: Witches' Thimble.

Woodland Sunflower
Helianthus divaricatus

◇◇◇◇◇◇◇◇◇◇◇◇◇◇◇◇

Growing Habits:

Mature Height: 6 feet
Mature Spread: 3 foot

Ontario Hardiness Zone: 3 to 7

Full Shade to Part Shade

Dry to Medium

Sand Loam, Loam

Helianthus divaricatus, or Woodland Sunflower, is a delightful addition to any Ontario garden, especially for those who prefer a low-maintenance approach. This native perennial, which typically grows to a height of 3-6 feet and spreads 2-4 feet, offers a stunning display of bright yellow flowers from summer to fall, with minimal effort required on your part. Its ability to thrive in part shade and dry conditions makes it perfect for those challenging spots in your yard. As a pollinator magnet, it's a boon for butterflies, bees, and birds. Not only does it beautify your garden, but it also supports local ecosystems. Its rhizomatous nature allows it to spread, gradually creating a naturalized, carefree look.

Woodland Sunflower is a versatile plant that is easy to grow in average, well-drained soils in partial shade. It is tolerant of various soil types, making it suitable for multiple garden settings. Whether you have a native/pollinator garden, naturalized area, or woodland garden, Woodland Sunflower can thrive in Ontario's climate. Once established, it is drought tolerant, making it a low-maintenance option for any garden. It is a high-value wildlife plant.

Neuroptera can benefit significantly from *Helianthus divaricatus*. While primarily carnivorous, some Neuroptera species supplement their diet with pollen, which *Helianthus divaricatus* abundantly provides. Additionally, the plant offers nectar, attracting nectar-feeding lacewings. Beyond food, the plant's structure can serve as shelter and hunting cover for these beneficial insects. By drawing in Neuroptera, *Helianthus divaricatus* can contribute to natural pest control as these insects prey on aphids, scale insects, and other garden pests.

Purple Coneflower
Echinacea purpurea

◇◇◇◇◇◇◇◇◇◇◇◇◇◇◇◇

Growing Habits:

Mature Height: 4 feet
Mature Spread: 1 foot

Ontario Hardiness Zone: 3 to 7

Part Shade to Full Sun

Dry to Medium

Loam

The purple coneflower, or *Echinacea purpurea*, is a remarkably adaptable and easy-to-care-for wildflower that flourishes in various soil types, including clay, loam, and sandy soils. It can thrive in dry and moderately moist soil conditions, making it an excellent choice for various garden settings. However, prolonged periods of drought can cause the plant to wilt, so it's best to provide regular watering during dry spells.

This striking wildflower is beautiful to pollinators, thanks to its vibrant and long-lasting blooms. As the season progresses, the large seed heads of the purple coneflower become a valuable food source for small birds, adding to its ecological importance. In addition to its ecological benefits, the purple coneflower has a rich history of medicinal use. Traditionally, it was employed in treating various ailments such as snake bites, headaches, and sore throats. It continues to be used for pharmaceutical purposes, particularly its potential respiratory relief properties.

While the Purple coneflower is officially classified as introduced to Ontario by VASCAN, it has gained widespread acceptance and usage in the region, leading many experts in the nursery industry in Southern Ontario to consider it almost native to the area. Its native range stretches from the northeastern United States to the border of Southern Ontario.

Bottlebrush Rye
Elymus hystrix

◇◇◇◇◇◇◇◇◇◇◇◇◇◇◇

Growing Habits:

Mature Height: 3 feet
Mature Spread: 2 feet

Ontario Hardiness Zone: 4 to 7

Full Shade to Full Sun

Dry to Medium

Sand Loam, Loam, Clay Loam

Bottlebush Rye is known for its remarkable ability to adapt to various shaded environments. It thrives in areas with dry shade or full sun as long as it receives enough moisture. Nonetheless, it tends to do best in partial shade conditions and well-drained soil. During the summer season, this grass produces long green inflorescences on stems that can reach up to three feet in height, resembling bottle brushes in appearance. As a cool-season perennial plant, it begins its growth cycle in the springtime.

Elymus hystrix is well-suited for woodland or rain gardens with partial shade. It is also an outstanding low-maintenance option for accentuating or bordering garden areas. The species derives its name from the Greek word "hystrix," which means hedgehog, a fitting descriptor for its bottlebrush-like inflorescence.

Bottlebrush rye is a summer smorgasbord for Hymenoptera. This flowering rye, with its abundance, attracts pollinators. The pollen-packed flowers provide a protein punch, the upright flower heads offer a multitude of flat landing platforms for accessible pollen and nectar collection, and the rye's extended bloom throughout the summer ensures a consistent food source during the Hymenoptera's active season. By providing both sustenance and a welcoming structure, bottlebrush rye becomes a vital resource for a variety of pollinators.

Indian Grass
Sorghastrum nutans

◇◇◇◇◇◇◇◇◇◇◇◇◇◇◇◇

Growing Habits:

Mature Height: 6 feet
Mature Spread: 2 feet

Ontario Hardiness Zone: 3 to 7

Part Shade to Full Sun

Dry to Medium

Sand Loam, Clay Loam

Yellow Indian grass, scientifically known as *Sorghastrum nutans*, is an impressive tall, clumping perennial grass that can reach heights ranging from 3 to 8 feet. It showcases wide, blue-green blades and bears a striking plume-like seed head that is soft and golden brown. This grass undergoes a breathtaking transformation in autumn, displaying deep orange to purple hues. With a somewhat metallic golden sheen, this warm-season grass is a pivotal component of tallgrass prairies and is highly valued by livestock for its nutritional value.

Yellow Indian grass is typically found in lowlands and tolerates occasional flooding and repeated burning. It often forms nearly pure stands in suitable habitats. During the fall, the plant's rich gold-and-purple sprays of flowers and seeds further enhance its visual appeal, making it an ideal ornamental addition to natural landscapes.

Interestingly, the foliage of yellow Indian grass plays a vital role in supporting the larvae of various skipper species, serves as a food source for birds, and provides nesting material in its dried state. Moreover, its persistent foliage throughout winter makes it an excellent option for creating shelter for wildlife, adding to its ecological significance.

By incorporating these native plants into your garden, you're creating a supportive environment for Neuroptera. These beneficial insects play a crucial role in controlling pest populations. Many Neuroptera species require specific plants for their life cycles, so a diverse selection is key. Remember, these delicate insects often prefer undisturbed areas, so consider creating small patches of wildflowers or leaving some areas of your garden to grow naturally. By providing the right conditions, you'll be rewarded with the sight of lacewings, green lacewings, and other fascinating Neuroptera making your garden their home.

Odonata

The Graceful Odonata of Ontario Gardens

◇◇◇◇◇◇◇◇◇◇◇◇◇

In the quiet corners of your garden, where gentle sunlight filters through the leaves and the sweet scent of blossoming flowers fills the air, a captivating dance unfolds among the Odonata. Picture a world where time stands still, and the shimmering wings of dragonflies and damselflies glisten in the sunlight, creating fleeting rainbows of colour.

These ancient insects, with a history spanning hundreds of millions of years, are not just fleeting beings but guardians of a timeless dance that transcends time constraints. Welcome to the enchanting realm of Odonata, where evolution has crafted a canvas of aerial mastery and predatory finesse.

In this enchanting display, dragonflies and damselflies emerge as winged marvels, capturing the imagination with their graceful flight patterns and vibrant colours. Yet, beyond their visual appeal, these insects serve as vital allies in maintaining the delicate balance of our local ecosystems, silently orchestrating a dance that supports the very fabric of life.

As you enter this world, you become part of a living tapestry woven by creatures whose existence predates the rise and fall of civilizations. Dragonflies, agile fliers with multifaceted eyes that grant them exceptional vision, and damselflies, delicate dancers with slender bodies and iridescent wings, coexist in a harmonious symphony that reverberates through the foliage of your garden.

This chapter delves into the mysteries of Odonata, shedding light on over one hundred and thirty species across Ontario, their incredible journey through the ages, and their intricate roles in the ecological drama of your garden. Join us as we uncover the secrets of these ancient aviators, exploring their physical beauty and profound impact on the interconnected web of life within the tranquil confines of your garden.

Autumn Meadowhawk,
Sympetrum vicinum

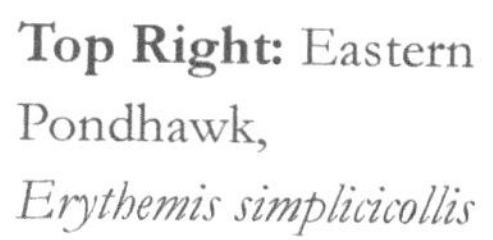

Top Right: Eastern Pondhawk, *Erythemis simplicicollis*

Top Left: Eastern Forktail, *Ischnura verticalis*

Bottom: Halloween Pennant, *Celithemis eponina*

Understanding Odonata

◇◇◇◇◇◇◇◇◇◇◇◇◇◇◇

Odonata, a captivating term derived from the Greek words "odon," meaning tooth, and "ata," signifying possessing, represents a fascinating group of predatory insects. These creatures are marked by their powerful mandibles, which resemble miniature teeth and are the key to their successful hunting strategies. This unique feature sets them apart and makes them a compelling subject of study.

Odonata's life cycle is a marvel in itself. It undergoes an incomplete metamorphosis, progressing through three distinct stages: egg, nymph (or naiad), and adult. The eggs, usually laid near or in water, hatch into nymphs. These nymphs, resembling their adult counterparts, go through several moults before emerging as fully-winged adults. This unique metamorphosis process is a critical characteristic that differentiates Odonata from other insects.

Delving into the taxonomic intricacies, the Order Odonata is a classification that encompasses two distinct suborders, each with unique characteristics and evolutionary adaptations:

1. **Dragonflies** (*Suborder Anisoptera*): The Odonata family's more prominent and robust members belong to the Suborder Anisoptera or dragonflies. With wings that differ in size and shape, these adept fliers showcase remarkable aerial prowess. Their multifaceted eyes, often covering a significant portion of their head, grant them an expansive field of vision, making them formidable hunters in the skies. Sitting dragonflies' wings are extended to each side of their body.

2. **Damselflies** (*Suborder Zygoptera*): The delicate damselflies, members of the Suborder Zygoptera, present a more slender and graceful demeanour. Their wings, usually of similar size and shape, epitomize elegance in flight. While not as robust as their dragonfly counterparts, damselflies have slender bodies to help execute their aerial finesse, darting through the air with an agility that belies their delicate appearance. Sitting damselflies fold their wings and hold them above their bodies.

As we delve into the world of Odonata, we uncover a realm where adaptation and specialization have shaped these insects into masters of the aerial domain. Beyond their enchanting beauty, comprehending the intricacies of these suborders deepens our admiration for the evolutionary processes that have crafted these extraordinary creatures over countless years.

Calico Pennant,
Celithemis elisa

Why Do We Need Odonata?

◇◇◇◇◇◇◇◇◇◇◇◇◇◇

The Order Odonata, which consists of dragonflies and damselflies, is one of the most overlooked but significant players in the complex tapestry of garden ecosystems. These enigmatic creatures are often the unsung heroes that maintain a healthy balance for thriving gardens. They serve as multi-faceted bio-control agents, pollinators and environmental health indicators.

Odonata are important in controlling pests in home gardens. Both adult dragonflies and damselflies eat insect pests like mosquitoes, aphids and flies. Their aerial stunts make them efficient predators, hence an organic alternative to chemical sprays. Through natural pest control, Odonata helps keep our gardens healthy and reduces the amount of human intervention required.

In addition to this, Odonata is a significant eco-pollinator. Dragonflies and damselflies may not be as widespread as bees or butterflies, but they are good at pollinating flowers by moving pollen grains while hunting prey. Flowering plants succeed in reproducing and establishing themselves more effectively, generating biodiversity and leading to resilience in the garden.

Furthermore, members of Odonata behave like bioindicators, revealing vital features about the overall well-being of garden ecosystems. They are susceptible to changes in water quality, habitat degradation and pollution. The number and diversity of Odonata species found within a garden can be a diagnostic tool for identifying other potential unknown environmental problems. By way of early warning signals about ecological imbalances, Odonata will always ensure that any form of gardening is sustainable in its activities, resulting in resilience.

The inclusion and preservation of Odonata in gardens are necessary for harmonious, thriving ecosystems.

Their contribution extends beyond visual aesthetics to Mosquito (*Family Culicidae*) control, pollination services, or even indicating the presence of toxic elements. As caretakers of nature, we must acknowledge and cherish Odonata's indispensability in our gardens' intricate rhythm of life.

Top: Chalk-fronted Corporal,
Ladona julia

Middle: Shadow Darner,
Aeshna umbrosa

Bottom: Dot-tailed Whiteface,
Leucorrhinia intacta

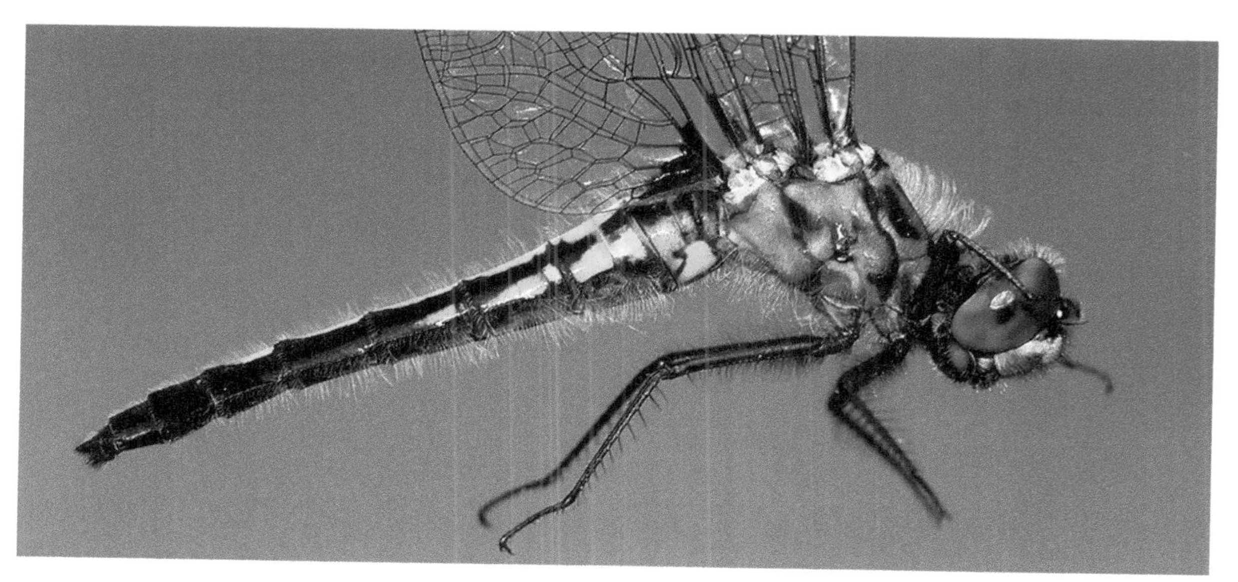

Morphological Marvels

Anatomy of Odonata

◇◇◇◇◇◇◇◇◇◇◇◇◇◇

We must observe the wings of Odonata. Dragonflies and Damselflies have two pairs of wings pointing at right angles, enabling them to fly exceptionally well. The venation pattern of their wings also tells us a lot about the different kinds within this Order. These insects' wings are designed to enable them to carry out aerial dances and maneuvers.

Odonata's eyes can also be examined as a strange part of its anatomy. Their big eyes have multiple facets, giving panoramic vision and allowing for accurate hunting and moving through space. Seeing almost everything around them simultaneously is crucial for survival as predators. Among other things, their visual and flight adaptations enable them to catch prey even while flying.

One of the most remarkable aspects of Odonata is their feeding apparatus, a tool for carnivory that is as efficient as it is fascinating. Their powerful jaws can capture and kill any prey, even in mid-air. This predatory prowess is a survival strategy and a boon for gardeners. By keeping garden pests in check, dragonflies and damselflies play a vital role in garden pest control, empowering gardeners in their quest for sustainable and pest-free crops.

Reproduction is another aspect of Odonatan biology that is highly interesting because many species have rather elaborate courtship rituals. It is not a surprise that dragonflies and damselflies perform aerobatics during mating or exhibit complex actions before getting into the copulation position itself. Once the eggs are laid by females either inside or near any water body, they hatch into aquatic larvae that moult numerous times before they metamorphose into adult form with four broad wings. Their larvae, adapted for life in water, promote ecological diversity in many aquatic habitats.

The study of Odonata anatomy is a journey through time, a testament to the evolutionary adaptations that have allowed these creatures to thrive for millions of years. The delicate-winged structure, the intricate compound eyes, the specially formed mouthparts, and the unique reproduction methods all contribute to the rich and functional diversity of these two groups within garden biodiversity. The morphological marvel of Odonata is a lesson in resilience and adaptation, inspiring gardeners to create environments that foster the thriving of such ancient and vital creatures. By recognizing the importance of dragonflies and damselflies in ecology, gardeners can contribute to the sustainability and resilience of their gardens and view these flying insects as integral parts of healthy ecosystems.

Eastern Pondhawk ,
Erythemis simplicicollis

Erythemis simplicicollis life cycle, egg, larva, adult.

Positives and Negatives
Welcoming Odonata in our Gardens

◇◇◇◇◇◇◇◇◇◇◇◇◇◇◇

Odonata, a term encompassing both Dragonflies and Damselflies, can make valuable contributions to gardens by feeding on pests. They are able hunters, as evidenced by studies that show a single dragonfly can eat hundreds of mosquitoes in one day. Their insatiable appetite for harmful insects also allows them to function as sustainable biological pest control agents in gardens, minimizing the need for chemical interventions.

Apart from controlling pests, Odonata also act as biomarkers indicating the state of health of their aquatic habitats. For instance, dragonflies lay their eggs in freshwater bodies, and the presence of their nymphs (or larvae) indicates clean and unpolluted water. The number and variety of dragonflies and damselflies in a garden can help judge the ecological status of water features.

Nevertheless, the main thing to consider is Odonata's indiscriminate predatory nature. Although they prey on pests, beneficial insects, including bees, butterflies, and other pollinators, may also be killed. Therefore, reaching a balance within the garden ecosystem is vital so Odonata can coexist with other key players in pollination and biodiversity.

The nymphs of these Dragonflies, which spend a considerable part of their early phase underwater, could impact smaller water organisms. In garden ponds, for instance, if they prey excessively upon other creatures dwelling therein, imbalances might result within these aquatic environments, requiring careful management to uphold a healthy, diverse water milieu.

Odonata contributes some benefits to gardens through natural pest control and is an environmental indicator. Nonetheless, gardeners should be aware of potential pitfalls, such as unintended effects on helpful insects or aquatic ecosystems, so that they can develop measures that promote balanced, thriving gardens.

Widow Skimmer,
Libellula luctuosa

Middle: Northern Spreadwing,
Lestes disjunctus

Bottom: Canada Darner, *Aeshna canadensis*

Discover Odonata

Ontario's Rich Odonata Diversity

To create an irresistible haven for beetles in your garden, strategically planting native Ontario species goes beyond mere horticulture—it's an invitation to witness a thriving, interconnected ecosystem. Here's a closer look at some native Ontario plants from Ontario Native Plants that we suggest you, as gardeners, plant and their role in enticing and supporting the diverse world of beetles and other wildlife.

By strategically integrating these native Ontario plants and others into your garden, you're not just adding to its visual appeal but crafting a vigorous stage where beetles play pivotal roles. This carefully curated selection becomes an essential component of your garden's biodiversity, supporting beneficial beetle species and enhancing the overall ecological resilience of your outdoor space. As these plants bloom and flourish, they beckon beetles, turning your garden into a living tapestry where each species contributes to the intricate dance of life, from the ground-dwelling protectors to the aerial pollinators. Your actions are crucial in this delicate balance.

Autumn Meadowhawk
Sympetrum vicinum

◇◇◇◇◇◇◇◇◇◇◇◇◇◇◇

Size: Adults range in size between 30 – 35 millimetres

Description Tips: The Autumn Meadowhawk dragonfly (*Sympetrum vicinum*) exhibits sexual dimorphism, with larger males displaying more vibrant colours than females. Both sexes lack distinctive markings, and immature individuals tend to be yellow or similar to females, gradually developing deeper coloration as they mature. In mature males, the face, eyes, thorax, and abdomen are shades of red to reddish-brown, while the legs range from yellow to reddish-brown. Conversely, females have a light brown face with brown and green eyes, a yellow to grey thorax, yellowish legs, and a brown abdomen. Females also have a prominent v-shaped ovipositor for egg-laying.

One of the most remarkable features of the Autumn Meadowhawk dragonfly is its ability to change colour in response to cool weather. In some regions, the male's vibrant red coloration shifts to orange, then brown as temperatures drop. This unique adaptation is a testament to the dragonfly's resilience and ability to adapt to its environment. The pterostigma, a coloured, thickened cell on the leading edge of each wing membrane near the tip, is a distinct reddish-brown. The larvae, known as 'nymphs,' measure 1.2-1.5 cm and exhibit mottled green and brown coloration. They have several large dorsal hooks along the abdomen, and the last two abdominal segments feature a single, large, rear-facing spine on each side.
Autumn Meadowhawks are expert fliers, capable of flying straight up and down, hovering like a helicopter, and even mating mid-air. They are diurnal and engage in active daytime flights.

These dragonflies are voracious predators, preying on small flying insects such as flies and mosquitoes. They rely on catching their prey while flying and are known for their quick movements, followed by moments of hovering and perching. Unlike most dragonflies, the *Sympetrum vicinum* exhibits unique courtship behaviour. They tend to rest at higher elevations away from water and complete most courtship rituals away from water bodies. They visit ponds or marshes only after mating, and the female is ready to lay an egg.

Common Whitetail
Plathemis lydia

◇◇◇◇◇◇◇◇◇◇◇◇◇◇◇

Size: Adults range in size between 38-48 millimetres

Description Tips: The Common whitetail is a widely distributed dragonfly species throughout much of North America. These dragonflies have unique features that make them easily identifiable.Males are easily recognizable by their striking appearance: a chunky white body contrasted by brownish-black bands on translucent wings. Females, on the other hand, have a brown body and a wing pattern similar to another species, the twelve-spotted skimmer. However, whitetail females can be distinguished by their smaller size, shorter body, and zigzag abdominal stripes. These dragonflies are adapted for swift flight, thanks to their broad wings and powerful muscles.

Plathemis lydia behaviour is their preference for low perching, often on the ground. Additionally, the whitetail larvae, also known as nymphs, are remarkable for their adaptability. They can thrive in waterways that are too degraded for most species, showcasing their resilience. These dragonflies can be found in natural aquatic habitats like streams, rivers, swamps, ponds and urban waters, demonstrating their ability to adapt to various environments.

As adults, *Plathemis lydia* are carnivorous, primarily feeding on mosquitoes, flies, and other small insects that they catch in flight. Their aerial acrobatics and sharp eyesight make them efficient hunters. The nymph stage is also carnivorous, feeding on various aquatic invertebrates.

Another exciting characteristic of whitetails is their preference for human spaces. They often perch on flat, sunny surfaces such as sidewalks, pathways, porches, balconies, picnic tables, boats, and even hats. This affinity for human spaces makes them a familiar and fascinating sight, enhancing our connection with these remarkable creatures.

Eastern Forktail
Ischnura verticalis

◇◇◇◇◇◇◇◇◇◇◇◇◇◇◇

Size: Adults range in size between 22-30 millimetres

Description Tips: The Eastern Forktail, scientifically known as *Ischnura verticalis*, gets its nickname "forktail" from the tiny projections on the tips of the males' abdomens, distinguishing the species. These damselflies have striking colours - vibrant lime green on the head, deep sky blue on the tail end, and a contrasting black hue. The head and thorax also feature some black striping, and the bulging eyes are dark on top and greenish below, with small green eyespots. The elongated, slender abdomen is primarily black but is highlighted by a stunning deep sky-blue colour on the last two segments at the tip. The females share a similar coloration with the males, though the final abdominal segments are even bluer.

Despite their stunning appearance and exciting behaviours, Eastern Forktails have relatively short adult lifespans, typically surviving for just a few months. During this time, feeding and mating behaviours take precedence. Adult males spend their time perching on branches or other objects, patrolling their territories, warding off rival males, and attempting to mate with females.

Although damselflies, including the Eastern Forktail, cannot sting, they may attempt to bite if handled, which can feel like a pinch. Not only are they fascinating creatures, but they also play a crucial role in ecosystems as beneficial predators. By preying on small flying insects during their adult stage, they help to control populations of harmful insects like mosquitoes.

In addition to aerial hunting, Eastern Forktail nymphs employ a highly effective strategy in their aquatic developmental stage. Resting quietly on submerged plants, they use their extendable, scoop-like jaws to snatch and draw in small aquatic animals swiftly. By regulating the populations of these small marine animals, they contribute to maintaining the balance of the ecosystem.

Ebony Jewelwing
Calopteryx maculata

◇◇◇◇◇◇◇◇◇◇◇◇◇◇◇

Size: Adults range in size between 39–57 millimetres

Description Tips: The Ebony jewelwing is a unique damselfly species, often found near streams and wooded areas. These creatures stand out with their iridescent green bodies and large black wings, creating a striking image of their natural habitat. When observing *Calopteryx maculata*, you may notice that the wings of younger adults are lighter and more transparent, while older individuals display more solidly black wings. Additionally, males typically have all-black wings and a brilliantly metallic blue-green body. On the other hand, females have a more subdued appearance with a greenish-bronze body and smoky-brown wings adorned with distinct white spots near the tips. The male's iridescent body exhibits a captivating colour shift, appearing either green or blue depending on the lighting conditions. The female's tiny white squares, known as stigmas, on the upper tips of her wings add charm to her appearance.

When a jewelwing takes flight from its perch along a stream, it's likely on a feeding mission. Although the insects it captures and consumes might be too small for human eyes, a good feeding perch is used for long periods. The damselfly also defends a territory of about six to ten feet along the stream from other damselflies, showcasing its territorial behaviour.

The female Ebony jewelwing requires plant material just below the water's surface to lay approximately 600 eggs. She accomplishes this by carefully dipping the tip of her abdomen underwater and using her ovipositor to deposit the eggs into a plant's stem or leaf. Meanwhile, the male diligently guards her, ensuring rival males are chased away. The nymphal offspring of the jewelwing embark on a remarkable journey. They spend their developmental stages in the aquatic environment of the stream, feeding on insect larvae and other invertebrates below the water's surface. In the early summer, they emerge from the water as beautiful-winged damselflies, a transformation that is nothing short of magical, evoking the image of woodland fairies taking flight.

Widow Skimmer
Libellula luctuosa

◇◇◇◇◇◇◇◇◇◇◇◇◇◇◇

Size: Adults range in size between 42 to 50 millimetres

Description Tips: The Widow skimmer, scientifically known as *Libellula luctuosa*, is a genuinely captivating dragonfly species. Its male counterpart boasts large, striking dark patches on the basal half of its wings, contrasting beautifully with its white shoulders and abdomen. The male's wings are adorned with distinct white and black bands, adding to its unique allure. In contrast, the female Widow skimmer is equally fascinating, with wings featuring brown tips, brown basal wing patches, and yellow stripes on its body. It's worth noting that the colours of this species deepen in richness as they mature, adding to their unique beauty.

The Widow skimmer is a dragonfly species many of us may have encountered without even realizing it. Its distinctive feature, the relative size of its wings, gives the impression of a larger body. This species is not confined to a specific region but is widespread, commonly observed throughout the United States and from Ottawa through Windsor in Ontario. It's a species that has adapted and thrived in various habitats, including ponds, lakes, marshes, and some slow streams, making it a familiar sight in many natural settings.

Observing *Libellula luctuosa* in its natural habitat is an enriching experience. Imagine spotting these dragonflies, their vibrant colours glinting in the sunlight, as they perch on the vegetated edges of local ponds or marshes. The best time for such an adventure is sunny days from June to September. Widow skimmers prefer tall plants for perching, so encouraging the growth of native grasses and wildflowers around neighbourhood waterways can provide ideal environments for their observation. Their preference for shallow, marshy pools is well-known, but they also have a unique affinity for more extensive meadows than other dragonfly species, making each sighting a special moment.

Odonates, with their vibrant colors and graceful flight, are a captivating addition to any garden. These beneficial insects are important predators of mosquitoes and other pests, making them invaluable allies in the fight against unwanted insects. By creating a suitable habitat with water features, emergent vegetation, and open spaces for flight, you can attract a diverse range of dragonflies and damselflies to your garden. Enjoy the mesmerizing spectacle of these aerial acrobats as they patrol your garden, providing both beauty and ecological balance.

Planting for Odonata

Discover Native Ontario Plants

Crafting a landscape and habitat that caters to the diverse needs of Odonata at different stages of life is essential to building a sanctuary that welcomes Odonata into the heart of your garden. Native Ontario plants, with their symbi-otic ties to the local ecosystem, provide the foundation for this enthralling synergy.

By incorporating these selected native plants into your garden and others, you not only enhance its aesthetic allure but also forge a haven that caters to the intricate needs of Odonata throughout their life cycles. This intentional culti-vation establishes a flourishing ecosystem where the enchanting dance of dragonflies and damselflies becomes an integral part of the symphony of nature, echoing through the verdant corners of your carefully nurtured haven. As gardeners, we are not just creators of beauty but stewards of nature, empowering us to shape the world around us.

Comprehending and appreciating Odonata's diverse world enhances our gardens' beauty and contributes to the in-tricate tapestry of our local ecosystems. As stewards of our natural spaces, let us embrace these enchanting in-sects, for in doing so, we welcome not just dragonflies and damselflies but graceful guardians who ensure the har-mony of our gardens, reminding us of the delicate balance of nature.

Red Osier Dogwood
Cornus sericea

◇◇◇◇◇◇◇◇◇◇◇◇◇◇◇◇

Growing Habits:

Mature Height: 6 to 13 feet

Ontario Hardiness Zone: 2 to 7

Part Shade to Full Sun

Medium to Wet

Sand Loam, Loam, Clay Loam, Organic

Cornus sericea, commonly known as Red osier dogwood or red-twig dogwood, is a stunning deciduous shrub prized for its year-round beauty and remarkable adaptability. Native to North America, this shrub shines with its vibrant red stems, offering a captivating display during the winter months when leaves have fallen.

Spring brings clusters of tiny white flowers, followed by white or bluish berries in summer that attract birds. In spring and summer, the oval-shaped and bright green leaves transform into a breathtaking palette of vibrant red, orange, or yellow in fall. This fast-growing shrub can reach heights between 6 and 13 feet.

For proper maintenance, it is advised to prune any old or dead branches during the spring season to promote the development of new branches. While it can adapt to various soil types, ranging from sandy to clay, it flourishes in soil rich in nutrients and consistently moist. Often used as a border shrub, it creates a remarkable visual contrast against an evergreen background.

The Red osier dogwood is a haven for Odonata thanks to its multi-faceted appeal. The shrub's abundant stems provide numerous perching sites, ideal for dragonflies and damselflies to rest and scan for prey or potential mates. In addition, the foliage and branches create a sheltered hunting ground, attracting flying insects that serve as a tasty meal for these aerial predators. Furthermore, in areas where Red osier dogwood grows near ponds or streams - prime breeding grounds for Odonata - the easy access to the water alongside suitable perching spots makes this plant particularly attractive. The Red osier dogwood becomes a valuable resource for Odonata populations by offering these essential elements.

Sweet Gale
Myrica gale

◇◇◇◇◇◇◇◇◇◇◇◇◇◇◇◇

Growing Habits:

Mature Height: 2 feet

Ontario Hardiness Zone: 2 to 7

Part Shade to Full Sun

Medium to Wet

Loam, Organic

Myrica gale, or Sweet gale, is a fascinating aromatic shrub that thrives in wet environments. Belonging to the Myricaceae family, this beautiful plant is characterized by its slender, fragrant leaves that exude a delightful aroma. Sweet Gale flourishes in moist, acidic soils, favouring habitats such as wetlands, bogs, and the banks of streams. Despite its modest size, typically reaching 3 to 6 feet, it significantly enhances the aesthetic appeal of various outdoor settings. Often used as a charming border plant around ponds, water features, footpaths, and property boundaries, the sweet gale never fails to capture the attention of admirers.

Sweet Gale, a plant that thrives in moist environments, plays a significant role in these ecosystems. It is commonly found in wetlands, bogs, marshes, lake margins, and even tolerates brackish water in the upper reaches of salt marshes and estuaries. One of its key roles is fixing nitrogen in the soil, a crucial ecological function. The flowers, though inconspicuous, appear in catkins before the leaves emerge in early spring. These flowers are followed by clusters of tiny winged nutlets encased in cone-like structures, which ripen in the fall. A fascinating fact about Sweet Gale is that both male and female flowers are borne on separate plants.

During the spring, before the emergence of its full foliage, the Sweet gale produces diminutive, dense, catkin-like flowers that release a delightful fragrance, making it an attractive haven for bees and other pollinators. Beyond its aesthetic and olfactory appeal, this native shrub is a vital habitat for birds and small wildlife. Its dense foliage provides an excellent source of protective cover.

Wood Poppy
Stylophorum diphyllum

◇◇◇◇◇◇◇◇◇◇◇◇◇◇◇◇

Growing Habits:

Mature Height: 2 feet
Mature Spread: 1 foot

Ontario Hardiness Zone: 4 to 7

Full Shade to Part Shade

Medium to Wet

Organic

The Wood poppy, also known as *Stylophorum diphyllum*, is a stunning perennial plant characterized by its gray-green, lobed and toothed leaves. The plant is well-regarded for its striking, large, poppy-like yellow flowers that bloom in small clusters atop leafy stalks. These flowers emerge from a stem that bears a pair of deeply lobed leaves, while additional leaves grow at the base of the plant. One particularly distinctive feature of the Wood poppy is its yellow sap.

This beautiful plant is an excellent addition to Eastern wildflower gardens, bringing a touch of natural elegance to any landscape. Notably, it is less invasive than introduced European species, making it a desirable choice for gardeners and conservationists. The name "*Stylophorum diphyllum*" is derived from Greek, meaning "two-leaved," a reference to the pair of opposite leaves under the flower.

In the wild, the Wood poppy is naturally distributed from Western Pennsylvania north to Wisconsin and Michigan and south to Arkansas, Tennessee, and southwestern Virginia. It also boasts isolated populations in northern Alabama and southern Ontario. However, due to its limited range in Ontario, the Wood poppy has been classified as an endangered species under Canada's SARA (*Species at Risk Act*) and by COSEWIC (*Committee on the Status of Endangered Wildlife in Canada*). This recognition underscores the significance of protecting and preserving this species for future generations

Switchgrass
Panicum virgatum

◇◇◇◇◇◇◇◇◇◇◇◇◇◇◇

Growing Habits:

Mature Height: 4 feet
Mature Spread: 3 feet

Ontario Hardiness Zone: 3 to 7

Full Sun | Dry to Medium | Sand Loam, Loam

Panicum virgatum, a hardy perennial bunchgrass native to North America, is not just a plant but a source of inspiration in any garden. It thrives in full sun and tolerates a wide range of soil moisture, from well-drained to moderately moist conditions. The slender, blue-green leaves, measuring 12-35 inches (30-90 cm) long, have a prominent midrib and exhibit a rolled appearance when young.

Switchgrass utilizes a C4 photosynthetic pathway, efficiently capturing carbon dioxide and thriving in hot, dry climates. During late summer (August-October), airy, reddish-purple flower panicles, measuring up to 24 inches (60 cm) long, emerge atop the foliage. These panicles contain tiny, single-flowered spikelets that eventually produce teardrop-shaped seeds. The seeds mature to a golden brown in fall and offer a valuable food source for songbirds and upland game birds.

With its deep root system extending several feet into the soil, Switchgrass is a low-maintenance yet highly beneficial addition to any landscape. This root system helps it access water during droughts and stabilizes soil, preventing erosion. While Switchgrass is generally low-maintenance, it benefits from being cut back to ground level in late winter or early spring before new growth emerges. This practice promotes bushier growth and removes dead plant material.

Blue Flag Iris
Iris versicolor

◇◇◇◇◇◇◇◇◇◇◇◇◇◇◇

Growing Habits:

Mature Height: 3 feet
Mature Spread: 2 feet

Ontario Hardiness Zone: 3 to 7

Part Shade to Full Sun

Medium to Wet

Loam, Organic

Iris versicolor, also commonly known as the blue flag, harlequin blue flag, larger blue flag, northern blue flag, and poison flag, is a flowering herbaceous perennial plant native to North America, specifically in the Eastern United States and Eastern Canada.

The unwinged, erect stems generally have basal leaves that are more than 1 cm (1⁄2 in) wide. Leaves are folded on the midribs to form an overlapping flat fan. The well-developed blue flower has six petals and sepals spread out nearly flat, and it has two forms. The longer sepals are hairless and have a greenish-yellow blotch at their base. The inferior ovary is bluntly angled.

Iris versicolor blooms in late spring with beautiful blue-purple flowers (to 4" wide) with bold purple veining. It prefers part shade to full sun and medium to wet soil. This plant is low maintenance and flourishes in sunny locations around ponds or other moist areas. It attracts many pollinators, including hummingbirds. However, be aware that the rhizomes of this plant are toxic to humans and animals if eaten.

Several factors make the Blue flag iris an excellent choice for attracting Odonata to your garden. Their tall, sword-shaped leaves provide ideal perching spots for adult dragonflies and damselflies. These insects need places to rest, bask in the sun, and hunt for prey, and the iris leaves offer the perfect platform. Additionally, the Blue flag iris thrives in wetland areas, which are the natural habitat for Odonata in their nymphal stage.

By incorporating these native plants into your garden, you're creating an ideal habitat for dragonflies and damselflies. These fascinating creatures play a crucial role in controlling insect populations and are a beautiful addition to any outdoor space. Remember, providing a combination of aquatic plants, emergent vegetation, and open areas is essential for supporting their life cycle. With careful planning and patience, your garden will become a vibrant haven for these aerial predators. Enjoy the enchanting spectacle of dragonflies and damselflies flitting among your Ontario native plants.

Orthoptera

The Symphony of Ontario's Orthoptera

◇◇◇◇◇◇◇◇◇◇◇◇◇◇

Step into the enchanting world of Orthoptera, where Ontario's vibrant meadows and woodlands teem with the captivating presence of crickets, grasshoppers, and katydids. Imagine a symphony of chirps harmonizing in the air, a grasshopper's graceful dance through the grass, and a katydid's remarkable camouflage amidst the foliage. With their finely honed adaptations, these incredible insects form a rich tapestry of biodiversity essential for Ontario's ecological harmony.

Delve deeper into the intricate world of Orthoptera and uncover the mesmerizing details that define these winged virtuosos. Explore the diverse species that inhabit the province, each contributing its unique melody to nature's grand symphony. From the musical courtship serenades that echo through the air to the remarkable leaps that define their movement, the behaviours of these insects are as fascinating as they are complex.

As we venture further, let's uncover the ecological significance of Ontario's over one hundred and fifteen Orthoptera species. These insects are not just performers but pivotal players in the captivating drama of predator and prey, pollinator and plant. Through a harmonious blend of scientific inquiry and boundless curiosity, this exploration reveals the vital roles these creatures play in ensuring the robustness and vitality of Ontario's ecosystems. Join us as we unravel the mysteries of Orthoptera, where captivating facts and sheer fascination converge in an exploration of the fascinating insect realm.

Roesel's Bush-Cricket,
Roeseliana roeselii

Top Left: Oak Bush-Cricket, *Meconema thalassinum*

Top Right: Pine Tree Spur-throat Grasshopper, *Melanoplus punctulatus*

Bottom: Northern Bush Katydid, *Scudderia septentrionalis*

What are Orthoptera

◇◇◇◇◇◇◇◇◇◇◇◇◇◇

Orthoptera, a taxonomic Order that includes crickets, grasshoppers, and katydids, is a fascinating and diverse group of insects. They are distinguished by their specialized hind legs, designed for prodigious leaps. This unique adaptation, a signature of the Order, gives them incredible agility, enabling them to perform spectacular evasive maneuvers and dynamic movements within their ecosystems.

Orthoptera comes from the Ancient Greek words "orthós," which means straight, and "pterá," which means wings. Orthoptera is an Order of insects that comprises grasshoppers, locusts, and crickets, including closely related insects such as bush crickets or katydids.

The Order is subdivided into two suborders:

1. **Caelifera** (*grasshoppers, locusts, and close relatives*): The Caelifera are a subdivision of orthopteran insects that occupy the Orthoptera subset. They include grasshoppers and grasshopper-like insects, as well as other superfamilies classified with them: members of the groups Tetrigoidea (the groundhoppers) or Tridactyloidea (the pygmy mole crickets).

2. **Ensifera** (*crickets and close relatives*): Ensifera is a suborder of insects that includes the various types of crickets and their allies, including true crickets, camel crickets, bush crickets or katydids, grigs, weta and Cooloola monsters.

Orthoptera undergoes incomplete metamorphosis, consisting of three stages: egg, nymph, and adult. The eggs hatch into nymphs, resembling miniature adults. Nymphs go through several moults, gradually developing wings and reproductive organs. Once reaching adulthood, Orthoptera retain their general form but with fully developed wings and reproductive capabilities.

Beyond their physical prowess, Orthoptera captivates with their enchanting auditory contributions. Crickets produce rhythmic chirps by rubbing specialized structures on their wings, creating a symphony that graces warm summer nights. Grasshoppers, too, emit distinctive sounds through various mechanisms, adding another layer to the aural landscape of their habitats. With their unique stridulatory apparatus, Katydids contribute their melodic calls to this captivating insect ensemble.

The evolutionary journey of Orthoptera has sculpted not only their jumping capabilities and sonic expressions but also bestowed upon them a mastery of camouflage. Their cryptic coloration and body structures allow them to seamlessly blend into their surroundings, becoming elusive inhabitants of diverse ecosystems.

This unique blend of evolutionary adaptations, from powerful hind legs to melodic calls and adept camouflage, positions Orthoptera as fascinating subjects for study and integral components of ecological systems. As we explore the depths of their characteristics, we unravel the biological marvels that make Orthoptera a captivating and indispensable ensemble within the insect kingdom.

Red-headed Bush Cricket,
Phyllopalpus pulchellus

Why Do We Need Orthoptera?

◇◇◇◇◇◇◇◇◇◇◇◇◇◇

Within garden ecosystems, the Order Orthoptera are the unsung heroes, their unique and irreplaceable contributions often overlooked. For instance, their feeding on organic material, including dead foliage and plants, accelerates the decomposition process, providing essential nutrients for the soil. They also serve as primary consumers in garden food chains, bridging the gap between plants and higher trophic levels. With their pivotal role, these insects maintain the delicate ecological balance of gardens, promoting biodiversity and adding a touch of life with their distinctive sounds. Understanding these specific contributions is critical to appreciating the health and functionality they bring to the garden.

Orthoptera are not just fascinating creatures but also practical allies in your garden. Their role as natural decomposers, feeding on organic material, including dead foliage and plants, accelerates the decomposition process, providing essential nutrients for the soil. This natural process of soil enrichment is a crucial advantage of having grasshoppers in your garden, as they contribute to its health and productivity. Understanding this can empower you to create a more sustainable garden ecosystem.

Orthoptera serve as primary consumers in garden food chains, bridging the gap between plants and higher trophic levels. While some may consider them pests due to their plant-eating habits, they are an essential part of a food chain that sustains a healthy garden ecosystem. Gardeners inadvertently foster other life forms by maintaining a diverse range of Orthoptera species, creating a more balanced and resilient ecosystem. This biodiversity is not just vital, but it's also a reward for the garden's health and sustainability, giving you confidence in your gardening practices.

Black-horned Tree Cricket,
Oecanthus nigricornis

Short-winged Meadow Katydid,
Conocephalus brevipennis

The sound of crickets or katydids is an essential indicator for gardeners to monitor within their gardens. Besides enhancing natural sounds heard outside structures, their rhythmic sound is used to measure how healthy an environment is. The absence or decline of these sounds may indicate changes in an ecosystem, which suggests that imbalances or stressors need to be sought out, such as fewer native plants for insects to eat. Spraying chemicals within your garden will also reduce most, if not all, insects, making Orthoptera species search elsewhere for the extra protein they need from dead insects.

Recognizing and appreciating the multifaceted contributions of Orthoptera is paramount for gardeners aiming to cultivate thriving and resilient ecosystems. These insects play a role in much more than just appearing physically present. They are involved in nutrient cycling, supporting biodiversity, and offering audible cues of ecological health, among other things. Incorporating various types of Orthoptera in gardens aims to foster a healthier community for all life within; thus, sustaining the intricate fragile web of life in a garden creates an environment where everything is harmonious and sustainable. To attract Orthoptera to your garden, consider planting a diverse range of native plants and providing suitable habitats such as log piles or meadow areas.

Black-legged Meadow Katydid,
Orchelimum nigripes

Morphological Marvels

Anatomy of Orthoptera

◇◇◇◇◇◇◇◇◇◇◇◇◇

One of Orthoptera's most distinguishing characteristics is its powerful hind legs, which are used for jumping. Athletic insects like grasshoppers can leap great distances due to bulky muscles on their hind legs and a unique mechanism at the tibiofemoral joint. This adaptation helps them escape predators and is critical in food finding and mate selection. The effectiveness of these hind legs demonstrates the evolutionary ingenuity that has allowed Orthoptera to survive in different environments.

Orthoptera wings are another morphological wonder with which they are endowed. Grasshoppers generally have two pairs of wings consisting of tegmina on the front, which are leathery sheaths that cover more delicate hindwings beneath them. The hindwings are membranous and expand during flight. On the other hand, crickets and katydids possess wings that lie flat against their body when not flying but can be unfolded immediately to take off or produce characteristic sounds like chirping, a vital manifestation in communication and mating patterns.

Orthoptera's mouthparts are modified for biting and chewing because they eat plants. For instance, locusts have formidable mandibles that enable them to chew through any foliage. Conversely, crickets and katydids display mandibles suited for eating plant matter and capturing smaller animals as prey because they exhibit omnivorous tendencies. This dietary adaptability places Orthoptera in critical positions in gardens, where they participate in pollination processes while checking pests.

Orthoptera has fascinating auditory organs. Crickets and katydids are known for producing elaborate sounds for communication, mainly through stridulation. The auditory organs on their forelegs help them detect these sounds, essential in courtship and territorial displays. These acoustic communications exhibit the complexity of Orthoptera behaviour and form part of their reproductive strategies.

Orthoptera's anatomy is a testament to various adaptations that allow these insects to survive in different ecological niches. From the strong hind legs and unique wing arrangement to the diversity of mouthparts and complicated auditory organs, grasshoppers, crickets, or katydids increase the biodiversity and functionality of gardens. Gardeners who understand the morphological marvels of Orthoptera can create habitats supporting these essential organisms. By growing lawns that acknowledge the ecological value of grasshoppers, crickets, or katydids, gardeners can practice sustainable gardening and ensure healthy gardens within thriving ecosystems.

Northern Bush Katydid,
Scudderia septentrionalis

Scudderia septentrionalis life cycle egg, nymph, and adult.

Positives and Negatives
Welcoming Orthoptera in our Gardens

◇◇◇◇◇◇◇◇◇◇◇◇◇◇◇

Garden ecological dynamics are significantly influenced by Orthoptera, which includes grasshoppers, crickets, and katydids. Grasshoppers are herbivores; hence, they play a significant role in nutrient cycling through the intake of plant materials, followed by the excretion of nutrient-rich frass, their waste product. This frass enhances soil fertility by returning nutrients to the soil. At the same time, crickets help to aerate soils through their burrowing activities, thus making it easier for water to pass through and distribute nutrients within it. Recycling nutrients and maintaining soil health in the garden ecosystem is a vital web of interactions involving Orthoptera.

Nevertheless, Orthoptera is associated with a high economic cost as a possible pest. For instance, grasshoppers can be agricultural pests that cause significant damage to crops, leading to financial losses for farmers. This is due to their high reproductive potential; hence, they can cause defoliation and loss in yield. These negative consequences should be minimized using integrated pest management measures such as habitat modification and safe biological controls. Similarly, crickets feed on various plant materials and, though helpful in air circulation within the soil, are problematic to farmers, especially when they are present in seed beds or young plantations.

Effective garden management requires a detailed understanding of Orthoptera's life cycle and behaviour patterns. Gardeners can use their knowledge of grasshoppers' inclination towards specific plants while selecting resistant varieties. Additionally, crops other than those growing in the garden, such as peppermint, lavender, and citronella, can help temporarily distract some crickets and grasshoppers from eating other crop plants or native plants.

In conclusion, Orthoptera contributes significantly to the ecological balance in gardens, mainly through participation in the cycles of nutrients and soil aeration by these organisms. Nevertheless, given their potential as pests, regulation must be practiced with informed decisions that limit negative implications while taking advantage of their positive impacts on the garden ecosystem.

Oblong-winged Katydid,
Amblycorypha oblongifolia

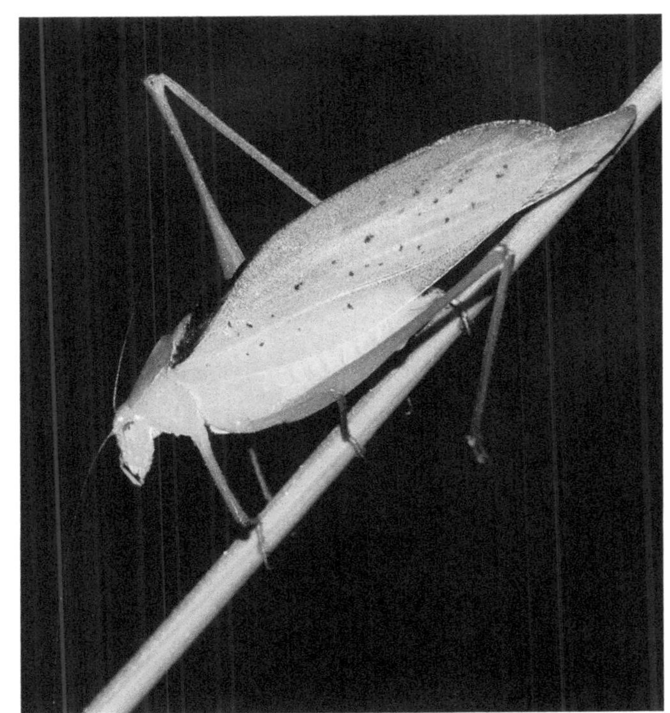

Middle: Wingless Mountain Grasshopper,
Booneacris glacialis

Bottom: Green-striped Grasshopper,
Chortophaga viridifasciata

Discover Orthoptera

Ontario's Rich Orthoptera Diversity

Ontario has a rich diversity of Orthoptera with over one hundred and fifteen species, each adapted to the province's varied landscapes. From dense forests to open meadows, you can encounter an array of crickets, grasshoppers, and katydids, showcasing the adaptability of these insects to different ecological niches.

Here Is a Closer Look At The Top Five Orthoptera In Ontario Gardens

Two-striped Grasshopper
Melanoplus bivittatus

◇◇◇◇◇◇◇◇◇◇◇◇◇◇◇

Size: Adults range in size between 26 to 40 millimetres

Description Tips: The Two-striped Grasshopper is a fascinating creature with varied colouring, ranging from brown to green, often with black or brown markings. One intriguing feature is the presence of two pale stripes extending from the grasshopper's eyes to the tips of its forewings. Additionally, a distinctive solid longitudinal black stripe can be observed on the hind legs of this species, making it a unique and exciting subject of study.

Ecologically, *Melanoplus bivittatus* are not just another insect. They are known to cause significant damage to cereal crops, particularly near field margins. The damage is primarily attributed to hatchling grasshoppers moving from their egg beds into the edges of fields. Unlike damage to grasslands, which tends to be evenly distributed, harm to cereal crops typically includes leaf notching and stripping. The most severe and costly damage occurs when grasshoppers sever stems just below the heads of maturing or mature crops, posing a severe threat to agricultural productivity.

During periods of elevated grasshopper populations and limited natural plant hosts, these insects display a voracious appetite. They will consume any available plants or products during their migrations for sustenance. This behaviour can have far-reaching implications for agricultural areas and the surrounding ecosystems, affecting crop yields and overall biodiversity. For instance, the loss of plant diversity due to their feeding habits can decrease the number of herbivorous insects, which in turn can impact the populations of insectivorous birds.

Carolina Locust
Plathemis lydia

◇◇◇◇◇◇◇◇◇◇◇◇◇◇◇

Size: Adults range in size between 32 – 58 millimetres

Description Tips: The Carolina Grasshopper, also known as *Dissosteira Carolina*, is a striking large-sized grasshopper endemic to North America and praised for its exceptional flying capabilities. When taking to the air, this species exhibits mesmerizing undulating and fluttering movements with sharp right-angle turns before gracefully descending to the ground. Typically, these grasshoppers prefer to fly at a moderate height, maintaining an altitude of 1 to 5 feet, and can traverse distances spanning anywhere from 3 to an impressive 70 feet. Their striking physical attributes include a body colour that ranges from a blend of gray-brown to reddish hues, adorned with delicate light speckling. Their distinctive black hind wings, bordered by a striking yellow margin, set them apart as the sole grasshopper species in Montana with this distinct feature, captivating the eyes of any observer.

Male Carolina Grasshoppers are known for their unique hovering flight displays over open terrain patches. These captivating aerial maneuvers can take them to heights ranging from 3 to 6 feet and, on rare occasions, even up to an awe-inspiring altitude of 15 feet. They can maintain this mid-air hovering stance for up to 15 seconds. It's important to note that such hovering flight performances are not commonly observed in other grasshopper species, making the opportunity to witness these displays a true testament to the patience and perseverance of the observer.

Interestingly, the *Dissosteira Carolina* has unique courtship rituals and encounters with rival males. It employs femur tipping and stridulation techniques, adding an intriguing behaviour layer unique to this extraordinary species. These behaviours are fascinating and serve as a testament to the complexity and diversity of nature.

Red-legged Grasshopper
Melanoplus femurrubrum

◇◇◇◇◇◇◇◇◇◇◇◇◇◇◇

Size: Adults range in size between 17- 30 millimetres

Description Tips: *Melanoplus femurrubrum* is commonly found in open habitats and displays a remarkable range of colours, including red-brown, yellow, dark brown, green, and olive green. One of their most striking features is their bright red or yellowish hind legs adorned with a black herringbone pattern. They belong to the spur-throated grasshoppers (*Genus Melanoplus*), characterized by a small pointy "spur" between their forelegs' bases. When startled, they can fly up to 40 feet at a fast and even pace, typically about a yard above the vegetation.

Regarding their ecological role, the Red-legged Grasshopper is an essential food source for wild game birds such as turkey and quail. Although these grasshoppers can be abundant, they can also harbour parasites that may be passed on to the consuming birds. Fortunately, these parasites typically do not cause severe harm unless the infestation is significant or the bird is weakened. Despite this risk, the presence of these grasshoppers is essential for the diverse bird population, as they play a crucial role in converting plant nutrients into a form suitable for consumption by birds and other insectivores.

These adaptable insects thrive in open areas with plenty of vegetation, making them familiar in fields, disturbed areas, and even near irrigation ditches. Beyond their jumping prowess, red-legged grasshoppers are known for their scientific fame. Their widespread presence and sensitivity to environmental changes make them invaluable model organisms for researchers studying climate impacts and insect behaviour. Plus, listen closely next time you spot one. You might hear their signature clicking sound, produced by the rhythmic rubbing of their hind legs.

Fall Field Cricket

Gryllus pennsylvanicus

◇◇◇◇◇◇◇◇◇◇◇◇◇◇◇◇

Size: Adults range in size between 15- 25 millimetres

Description Tips: The Fall Field Cricket, scientifically known as *Gryllus pennsylvanicus*, stands out with its shiny black exoskeleton and rounded head. Its tegmina, or forewings, display colours from light brown to black, showcasing fascinating variations within the species. Notably, female crickets, with their slightly brown wings and a prominent ovipositor that extends beyond their body length, are more extensive than their male counterparts. The closer positioning of female antennae, in contrast to the spread-out male antennae, is another unique feature.

Gryllus pennsylvanicus is a nocturnal insect with a taste for plants and small insects. These black or dark brown critters favour grassy areas, burrowing in soil found in fields, forest edges, or even near your house! More significant than some cricket cousins, they reach adulthood after 8-9 immature stages, emerging in late July or early August. Their one-season lifespan revolves around finding food at night and attracting mates with chirping songs in late summer and fall. While beneficial as pest controllers sometimes, these solitary insects can become a nuisance when they find their way indoors.

Beyond their physical traits, the ecological importance of these crickets is equally fascinating. Their songs have been the subject of extensive scientific study, revealing characteristics. For instance, by counting the number of chirps a male cricket produces in 13 seconds and adding 40 to this number, one can estimate the local temperature in degrees Fahrenheit. This intricate relationship between their songs and the environment is a testament to the complexity of nature. Furthermore, Fall Field Crickets play a crucial role in the ecosystem as a food source for numerous bird species. As herbivorous insects, they contribute to the conversion of plant nutrients into a form that can be consumed by birds and other insectivores, thereby supporting the delicate balance of the ecosystem.

Drumming Katydid
Meconema thalassinum

◇◇◇◇◇◇◇◇◇◇◇◇◇◇◇◇

Size: Adults range in size between 14-19 millimetres.

Description Tips: The Drumming katydid, also known as the Oak bush cricket, is a captivating little insect native to Europe. It has a surprisingly established presence in parts of eastern North America. This lime-green marvel belongs to the Tettigoniidae family, sharing kinship with katydids and grasshoppers.

The Drumming katydid is a fascinating insect with vibrant green colouring and a contrasting yellow-orange stripe running along its back. Females can be identified by their long, pointed ovipositor, which is used for laying eggs, while males possess curved cerci resembling antennae at their rear ends.

These little predators prefer to inhabit deciduous trees and surrounding vegetation, particularly in woodlands and forested areas. They are most active in late summer and fall, expertly hunting small insects like aphids with sharp eyesight and agility. Unlike their katydid relatives, who utilize wings for sound production, oak bush crickets have a unique drumming technique. This drumming, which they create by rapidly tapping a hind leg on a leaf, serves multiple purposes. It helps them attract mates, defend their territory, and communicate with other species.

Meconema thalassinum life cycle revolves around one generation per year. Adults emerge in late summer and remain active until fall. During this time, females lay their eggs in bark crevices, where they overwinter until spring. Once hatched, the nymphs undergo several moulting stages before adulthood in late summer.

An interesting yet concerning fact about these insects is their vulnerability to a parasitic worm called Spinochordodes tellinii. This parasite can manipulate the cricket's behaviour, making it more drawn to water, an essential element for the parasite's life cycle. Additionally, despite their introduction to North America, the range of Drumming Katydid is still expanding.

Orthoptera, including grasshoppers, crickets, and katydids, are often overlooked but play important roles in the garden ecosystem. While some species can be considered pests, many are beneficial, serving as food sources for birds and other wildlife. By understanding their life cycles and habitat preferences, you can create a garden that supports a balanced Orthoptera population. Remember, providing a diversity of plants, including native grasses and wildflowers, can attract a variety of these fascinating insects.

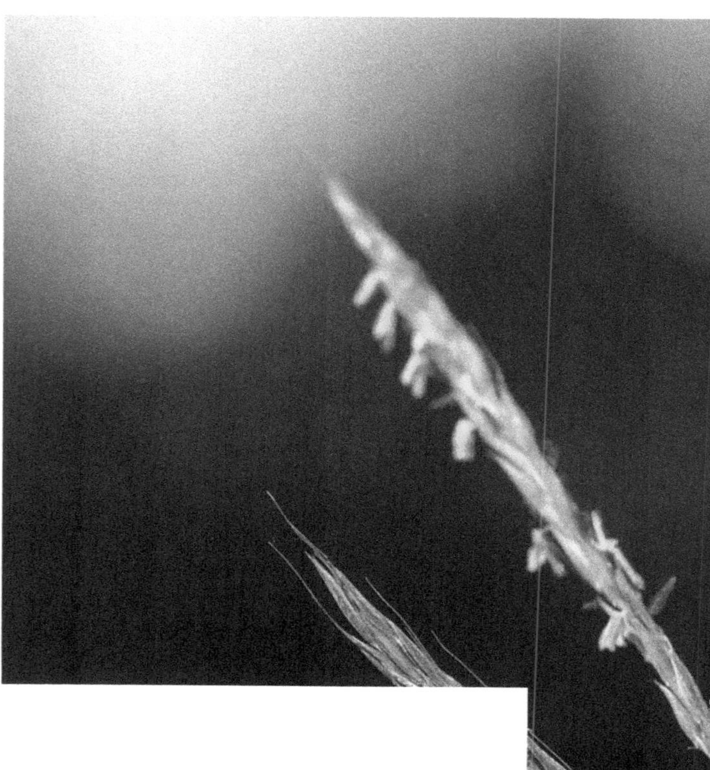

Planting for Orthoptera

Discover Native Ontario Plants

If you want to attract Orthoptera to your garden, it is essential to choose native Ontario species. These plants have evolved to thrive in the local environment and provide the best resources for the region's native fauna, including Orthoptera. Using Ontario Native Plants, we recommend several native Ontario plants to help sup-port these insects and other wildlife throughout their life cycles.

It is crucial to embrace the diversity and understand the ecological functions of Orthoptera species within our garden ecosystems. Grasshoppers, for instance, are herbivorous insects that can impact plant communities by influencing vegetation dynamics and nutrient cycling. Crickets play essential roles in decomposition through their feeding habits on organic matter, contributing to the breakdown of dead plant material and nutrient re-lease into the soil. Moreover, the interactions between Orthoptera and plants go beyond mere consumption. Certain species are known pollinators, facilitating the reproduction of flowering plants and contributing to genetic diversity. In this intricate dance of coexistence, unseen performers, such as katydids, with their cryptic appearances and nocturnal activities, become indispensable contributors to the flourishing of life within our gardens.

Therefore, by recognizing and respecting the presence of Orthoptera in our gardens, we encourage a deeper connection with the natural world. These insects transform our gardens into dynamic ecosystems, where their often-overlooked roles become vital threads in the intricate web of life. This understanding gives us insight into the delicate balance that sustains our shared environment's biodiversity and ecological stability.

Blue Eyed Grass
Sisyrinchium montanum

◇◇◇◇◇◇◇◇◇◇◇◇◇◇◇

Growing Habits:

Mature Height: 1.5 feet
Mature Spread: 0.5 feet

Ontario Hardiness Zone: 4 to 7

Part Shade to Full Sun

Dry to Medium to Wet

Sand Loam

Blue-eyed grass, also known as *Sisyrinchium angustifolium*, belongs to the iris family and is a delightful addition to any garden. Despite its name, this fast-growing and drought-tolerant wildflower is not an actual grass. It displays elegant six-petaled violet flowers adorned with a striking yellow "eye" at the center, which can appear singly or in charming clusters atop long stalks. These lovely blooms grace the garden from spring well into early summer.

Blue-eyed grass thrives in dry, exposed roadsides and well-drained, grassy meadows. It can also be successfully cultivated in rock gardens, along pathways, or as a colourful feature at the front of garden beds. This plant is beautiful to butterflies, flies, and bees, making it a beneficial addition to any pollinator garden.

In addition to its ornamental value, Blue Eyed Grass boasts culinary potential. Its tender leaves and delightful flowers are edible and can be incorporated into fresh salads, flavorful soups, and hearty stews. This versatile wildflower enhances the garden's aesthetic appeal and offers a delightful culinary experience for those who appreciate its edible qualities.

Heartleaf Foam Flower

Tiarella cordifolia

◇◇◇◇◇◇◇◇◇◇◇◇◇◇◇

Growing Habits:

Mature Height: 1 foot
Mature Spread: 2 feet

Ontario Hardiness Zone: 4 to 7

Full Shade to Part Shade

Medium

Loam, Clay Loam

The Heartleaf foamflower, scientifically known *Tiarella cordifolia* is a delightful perennial wildflower native to the Eastern United States and Canada. It is beloved for its graceful, long, slender stamens that give it a bubbly and ethereal appearance. The plant produces beautiful spikes of petite, star-shaped white flowers that grow in compact racemes on stalks ranging from 6 to 16 inches in height. These elegant flowering spikes emerge above a dense cluster of attractive, deeply lobed leaves.

The flower stalk is leafless in the northern regions of its range, while in the South, it is adorned with charming heart-shaped leaves. As the plant matures, it sends out runners, gradually forming expansive colonies through its underground spreading stems, making it an excellent choice for groundcover in shady, wooded areas.

The origin of the Greek genus name "tiarella" can be traced back to the resemblance of the plant's pistil to a turban worn by the Persians. The delicate, feathery, elongated, terminal clusters of small, white to pinkish flowers, in combination with the delicate texture of the stamens, give this plant its common name, foamflower. This unique and lovely wildflower thrives in woodland settings and adds a touch of elegance to any garden.

Prairie Dropseed
Sporobolus heterolepis

◇◇◇◇◇◇◇◇◇◇◇◇◇◇◇

Growing Habits:

Mature Height: 2 feet
Mature Spread: 2 feet

Ontario Hardiness Zone: 3 to 7

Full Sun

Dry to Medium

Sand Loam, Clay Loam

Prairie Dropseed, scientifically known as *Sporobolus heterolepis*, is a long-lived, warm-season, perennial bunchgrass native to the prairies of central North America. This grass is highly valued for its striking ornamental features and adaptability to various soil types and environmental conditions. Its narrow, arching foliage forms dense, finely textured clumps, providing a unique visual appeal to landscapes. In addition to its attractive appearance, Prairie Dropseed also releases a pleasant, cilantro-like fragrance when its leaves are brushed or crushed, adding an extra sensory dimension to outdoor spaces.

Furthermore, this grass is resilient in hot, dry climates, making it an ideal choice for xeriscaping and water-wise gardening. Despite its preference for well-drained soils, Prairie Dropseed can thrive in various soil types, including dry, rocky terrain and heavy clays, offering versatility to gardeners facing challenging growing environments.

Another highlight is *Sporobolus heterolepis's* low-maintenance nature, which does not require frequent watering or extensive upkeep once established. Although it produces seed heads in the fall, it generally does not self-seed aggressively, making it a reliable ground cover option for low-maintenance gardening and landscaping projects. Its ability to withstand heat, drought, and poor soil conditions, as well as its delightful fragrance and ornamental appeal, solidifies Prairie Dropseed's status as an exceptional choice for sustainable and visually captivating landscapes.

Big Bluestem
Andropogon gerardii

◇◇◇◇◇◇◇◇◇◇◇◇◇◇

Growing Habits:

Mature Height: 7 feet
Mature Spread: 2 feet

Ontario Hardiness Zone: 3 to 7

Part Shade to Full Sun

Dry to Medium

Sand Loam

Big Bluestem, scientifically known as *Andropogon gerardii*, is a towering warm-season bunchgrass that plays a crucial role in Ontario's tallgrass prairie ecosystem. Its remarkable resilience stems from its extensive root system, which can delve over 10 feet below the soil surface. This feature allows it to withstand harsh environmental conditions.

Beyond its impressive fortitude, Big Bluestem is highly esteemed for contributing to the structure and beauty of prairie and meadow landscapes. During winter, its striking bronze-coloured foliage adds visual allure to the surroundings. It is also well-suited for rain gardens, prairie gardens, and naturalized areas, where it can thrive and enhance the natural beauty of these settings.

In addition to being a visual delight, *Andropogon gerardii* serves essential ecological functions. Its robust stems endure through winter, providing sustenance for various bird species. Furthermore, the leaves of Big Bluestem support numerous caterpillar species, thus contributing significantly to the overall biodiversity of the ecosystem. This further emphasizes Big Bluestem's invaluable role in supporting the intricate web of life within the prairie ecosystem.

Blue Stem Goldenrod
Solidago caesia

◇◇◇◇◇◇◇◇◇◇◇◇◇◇◇◇◇

Growing Habits:

Mature Height: 3 feet
Mature Spread: 1 feet

Ontario Hardiness Zone: 4 to 7

Full Shade to Part Shade

Dry to Medium

Loam, Clay Loam

The Blue Stemmed Goldenrod stands out with its unique features, particularly its non-aggressive spreading. Unlike some goldenrods, it does not spread aggressively, providing reassurance about its growth in your garden. Its graceful arching stems, adorned with hundreds of small yellow flowers, are a sight to behold. The distinct purplish colour of the stems adds to its allure. This native plant, found in the rich woodlands, thrives in light shade but can also tolerate full sun. A mature clump of this plant is a striking addition to any semi-shade garden, brightening it up late in the season.

Solidago caesia is an attractive goldenrod for various garden styles: open woodland garden, border, cottage or butterfly garden. As with all goldenrods, it is a desirable source of late-season pollen. Other common names include Wreath Goldenrod. However, it's important to note that it can become invasive in certain conditions, such as in rich, moist soils. Therefore, it's best to plant it in well-drained soil to prevent this issue.

Blue-stem goldenrod is a haven for Orthoptera, serving as a vital food source for these insects. This beneficial plant provides a two-fold advantage. First, it offers a nutritious feast, with its leaves and flower parts serving as a food source. Second, the tall, straight stems and dense foliage create a protective shelter. This allows orthoptera to hide from predators and harsh weather conditions, promoting healthy populations.

By incorporating these native plants into your garden, you're creating a thriving habitat for Orthoptera. These insects play a vital role in the food chain, serving as prey for birds and other wildlife. Remember, a diverse plant community will attract a variety of Orthoptera species. From the vibrant colors of grasshoppers to the enchanting songs of crickets, these insects add a unique charm to your garden. Enjoy the symphony of nature as you watch these fascinating creatures thrive among your native plants.

In Conclusion

◇◇◇◇◇◇◇◇◇◇◇◇◇◇◇

The delicate balance of our garden ecosystems relies on the presence of native plants and insects in Ontario. These native plants are vital for sustaining the diversity within our gardens and ensuring the well-being of our local wildlife. Among the fascinating insect orders that contribute to this intricate dance of life in our garden ecosystems are Coleoptera, Diptera, Hemiptera, Hymenoptera, Lepidoptera, Neuroptera, Odonata, and Orthoptera.

Native plants are the unsung heroes that support these buzzing partners. The decline in insect populations can be partly attributed to the loss of their natural habitats due to invasive species. Gardeners must recognize the essential relationship between animals and vegetation. By incorporating plants indigenous to our local areas, we can help ensure the survival of vital insects, leading to more resilient and vibrant gardens.

The connection between the top eight insect orders and native plants is vital for any gardener aiming to cultivate a thriving, sustainable garden. For example, incorporating native plants like coneflowers and goldenrods can encourage a rich diversity of beetles, providing food and aesthetic appeal to your garden. Similarly, native flowers such as Wild lupine and New England aster can attract beneficial flies, offering pollination benefits and natural pest control.

Cultivating native blooms like the Buttonbush and Thimbleberry can support bees' buzzing brilliance, enhancing your garden's beauty and productivity. Additionally, introducing native wildflowers such as butterfly milkweed and asters can sustain butterflies' delicate dance, creating a sanctuary for caterpillars and adult butterflies. Similarly, incorporating native grasses like switchgrass can provide shelter and sustenance for delicate insect predators such as lacewings

Two-marked Treehopper Complex,
Complex Enchenopa binotata

Incorporating native aquatic plants like Wood Poppies can attract the intriguing world of dragonflies while enhancing water features within your garden. Cultivating native grasses like Little Bluestem and Prairie Dropseed can create the perfect habitat for grasshoppers, embracing the biodiversity within your garden's greenery. It's essential to research the native plants in your region while supporting Ontario Native Plants for all your native plant needs across Ontario.

As we admire the beauty and vast diversity of Ontario's insect world in our gardens, we must recognize the crucial role that native plants play in maintaining this delicate balance. By planting native flora, we become caretakers of a thriving ecosystem that amplifies the hum of life and enhances the charm of our gardens. As we witness a decline in the insect population, it's time to act and transform our gardens into sanctuaries that resonate with these extraordinary creatures' wings, buzz, and songs. This will ensure a balanced coexistence between nature and the curated spaces we call our own.

This guide aims to help you familiarize yourself with the eight most prevalent insect orders and the native plants that can attract them. It's designed to assist you in identifying these eight standard orders and developing an appreciation for native plants and insect species. Using the insights from this guide and others, we can collectively contribute to the proliferation of indigenous plant species and the augmentation of insect diversity across Ontario, Canada, and the world.

Wild Geranium,
Geranium maculatum